SpringerBriefs in Quantitative Finance

More information about this series at http://www.springer.com/series/8784

Tim Leung • Marco Santoli

Leveraged Exchange-Traded Funds

Price Dynamics and Options Valuation

 Springer

Tim Leung
Columbia University
New York City, NY, USA

Marco Santoli
Columbia University
New York City, NY, USA

ISSN 2192-7006   ISSN 2192-7014   (electronic)
SpringerBriefs in Quantitative Finance
ISBN 978-3-319-29092-8   ISBN 978-3-319-29094-2   (eBook)
DOI 10.1007/978-3-319-29094-2

Library of Congress Control Number: 2016930168

Mathematics Subject Classification (2010): 91G10, 91G20, 91G70, 91G80, 62F30

Springer Cham Heidelberg New York Dordrecht London

Printed on acid-free paper

Springer International Publishing AG Switzerland is part of Springer Science+Business Media (www.
springer.com)

To Kelly and Vivian

Preface

Leveraged exchange-traded funds (LETFs) are relatively new financial products liquidly traded on major exchanges. They have gained popularity with a rapidly growing aggregate assets under management (AUM) in recent years. Furthermore, there are now derivatives written based on LETFs. This book aims to provide an overview of the major characteristics of LETFs, examine their price dynamics, and analyze the mathematical problems that arise from trading LETFs and pricing options written on these funds.

When writing this book, we aim to make it useful not only for graduate and advanced undergraduate students but also for researchers interested in financial engineering, as well as practitioners who specialize in trading leveraged or non-leveraged ETFs and related derivatives.

In the first part of the book, we assume very little background in probability and statistics in our discussion of the price dynamics of LETFs. Nevertheless, new insights and trading strategies are discussed with mathematical justification and illustrated with a host of examples using empirical data. Our emphasis is on the risk analyses of LETFs and associated trading strategies. The second part focuses on the risk measurement for LETFs, and we provide a number of formulas for instant implementation. In the final part, we present the analytical and empirical studies on the pricing and returns of options written on LETFs. Our main objective is to examine a consistent pricing approach applied to all LETFs. This allows us to identify any price discrepancies across the LETF options markets. As the market of ETFs continues to grow in terms of market capitalization and product diversity, there are plenty of new problems for future research. In the final chapter, we point out a number of new directions.

We would like to express our gratitude to several people who have helped make this book project possible. Parts of the book are based on the thesis of my Ph.D. student and coauthor, Marco Santoli, who has been funded by the Department of Industrial Engineering and Operations Research at Columbia University throughout his doctoral study. Various chapters have been used in several courses at Columbia University and have benefited from students' feedback and questions. Several Columbia Ph.D. and master's students, who participated in exploratory projects on ETFs, have also helped shape the materials.

We greatly appreciate the helpful remarks and suggestions by Carol Alexander, Rene Carmona, Peter Carr, Alvaro Cartea, Michael Coulon, Emanuel Derman, Jean-Pierre Fouque, Paul Glasserman, Paolo Guasoni, Sam Howison, Sebastian Jaimungal, Ioannis Karatzas, Steven Kou, Roger Lee, Vadim Linetsky, Matt Lorig, Mike Ludkovski, Andrew Papanicolaou, Ronnie Sircar, Charles Tapiero, Agnes Tourin, Nizar Touzi, and Thaleia Zariphopoulou, as well as the ETF tutorial participants at the Risk USA Workshop 2014 and the INFORMS Annual Meeting 2015. In addition, we are grateful for the constructive comments from four anonymous referees and the series editor during the revision of the manuscript. Lastly, we thank Donna Chernyk of Springer, USA, for encouraging us to pursue this book project.

New York City, NY, USA   Tim Leung
Thanksgiving Day, 2015

Contents

Chapter 1

Introduction

The market of exchange-traded funds (ETFs) has grown substantially in recent years. In 2015, the US ETF industry consists of 1,600 funds with over \$2.1 trillion in assets under management (AUM).[1] ETFs are traded on major exchanges, such as NASDAQ and NYSE, like stocks. An ETF is an investment fund, but in contrast to other index funds or mutual funds, most ETFs are designed to track a prespecified index, such as a stock index, bond index, commodity index, volatility index, etc., while charging a small expense fee over time. They offer investors access to trade in different sectors and asset classes, with domestic or international exposure. Equity ETFs constitute the majority of AUM, with over \$1.5 trillion, followed by fixed income and commodity ETFs, with over \$320 billion and \$53 billion, respectively.

Figure 1.1 depicts how the ETF market works. First, there is the primary market where ETFs are created or redeemed through the interactions between the ETF provider and authorized participants (APs). The APs in the primary markets include ETF managers representing an ETF provider like ProShares, authorized dealers, and market makers. Authorized dealers can give ETF units to the ETF manager and receive the underlying assets in return. In this redemption process, ETFs units are reduced in the primary market. In reverse, if the authorized participants give the ETF manager a physical basket of the underlying assets (in-kind transfer), then the ETF manager will create and deliver units of the ETF of the same cash value. This creation process

---

[1] See "Should You Fear the ETF?" Wall Street Journal, December 6, 2015. Available at http://www.wsj.com/articles/should-you-fear-the-etf-1449457201.

© The Author(s) 2016
T. Leung, M. Santoli, *Leveraged Exchange-Traded Funds*, SpringerBriefs in Quantitative Finance, DOI 10.1007/978-3-319-29094-2_1

effectively increases the number of units of the ETF. The exchanges where the ETFs are liquidly traded are the secondary market where buyers obtain shares from existing ETF holders.

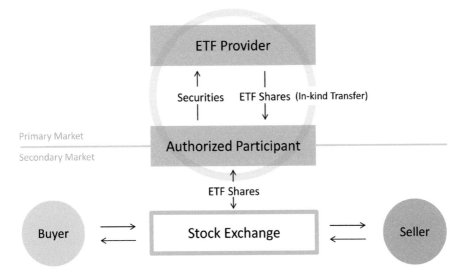

Fig. 1.1: ETF market: ETF redemption allows the authorized participants to give ETF units to the ETF provider and receive the underlying assets in return. In reverse, if the authorized participants give the ETF provider a physical basket of the underlying assets (in-kind transfer), then the ETF provider will create and deliver units of the ETF of the same cash value. The exchanges where the ETFs are liquidly traded are the secondary market where buyers obtain shares from existing ETF holders.

Within the ETF market, leveraged ETFs (LETFs) have gained popularity among investors with over \$61 billion AUM as of October 31, 2014.[2] LETFs are typically designed to replicate multiples of the daily returns of some underlying index or benchmark. For example, the ProShares Ultra S&P 500 (SSO) is supposed to generate twice the daily returns of the S&P 500 index, minus a small expense fee. Moreover, investors can take a bearish position on the underlying index by longing an inverse LETF (with a negative leverage ratio) without a margin account. An example is the ProShares UltraShort S&P 500 (SDS) on the S&P 500 with leverage ratio of -2. In addition,

---

[2] "Short & Leveraged ETFs/ETPs Global Flows Report," Boost ETP, a WisdomTree Company.

both long and short triple LETFs are also available for various underlyings. For many investors, LETFs are a highly accessible and liquid instrument. They also tend to be more effective during periods of large momentum and low volatility.

On the other hand, LETFs have drawn a number of criticisms. Some argue that they tend to underperform over extended (quarterly or annual) investment horizons, as compared to the promised multiple of the underlying index returns (see Figures 2.2–2.3 below). The underperformance has sometimes been attributed to ill-timed rebalancing, returns compounding, and the use of derivatives to replicate returns. Cheng and Madhavan (2009) illustrate that the LETF value can erode over time due to its dependence on the realized variance of the underlying index, coupled with daily rebalancing. Avellaneda and Zhang (2010) also discuss the path-dependent performance and potential tracking errors of LETFs under both discrete-time and continuous-time frameworks. In fact, SEC has issued an alert announcement regarding the riskiness of LETFs,[3] and investigated whether LETFs would create a feedback effect and lead to increased market volatility.[4]

For LETF holders and potential investors, it is very important to understand the price dynamics and the impacts of leverage ratio on the risk and return of each LETF. Therefore, we begin with a series of empirical studies on the returns and tracking performance of LETFs in Chapter 2. A number of market observations suggest that LETFs suffer from value erosion that is proportional to the volatility of the reference index and holding horizon, and it is more severe for highly leveraged ETFs. We present a stochastic framework to model the evolutions of the reference index as well as the LETFs. In turn, we can show mathematically how the return of an LETF is related to that of the reference. From this we identify precisely the roles of the leverage ratio, holding horizon, and realized variance of the reference index on the LETF's return over time.

Chapter 2 motivates the analysis of the risk implications of leverage ratio. Hence, in Chapter 3, we provide a quantitative risk analysis of LETFs with an emphasis on the impact of leverage and investment horizon. Given an investment horizon, different leverage ratios imply different levels of risk. Therefore, we introduce the idea of an admissible range of leverage ratios. These are the

---

[3] See the SEC alert on http://www.sec.gov/investor/pubs/leveragedetfs-alert.htm.

[4] See "SEC Looks Into Effect of ETFs on Market," Wall Street Journal, September 7, 2011.

leverage ratios for which the associated LETFs satisfy a given risk constraint based on, for example, the value-at-risk (VaR) and conditional VaR. This idea can help investors exclude LETFs that are deemed too risky. Moreover, we discuss the concept of admissible risk horizon introduced in Leung and Santoli (2012), which allows the investor to control risk exposure by selecting an appropriate holding period. In addition, we compute the intra-horizon risk and find that higher leverage can significantly increase the probability of the LETF value hitting a lower level. This leads us to evaluate a stop-loss/take-profit strategy for LETFs. In particular, we determine the optimal take-profit level given a stop-loss risk constraint. Lastly, we investigate the impact of volatility exposure on the returns of different LETF portfolios.

In Chapter 4, we turn to study options written on LETFs. We begin by conducting an empirical analysis on the returns of S&P 500 based LETF options. We also compare option prices across leverage ratios by analyzing the so-called *implied dividend* ratio – computed using the call-put parity relation. The observed discrepancies can be interpreted as the result of different market frictions (i.e., borrowing costs) present across leverage ratios.

Since LETFs with the same reference index ultimately share one common source of risk, it is natural to wonder how the prices of LETF options are related. Moreover, it is of both theoretical and practical importance to develop a pricing framework that generate LETF option prices that are arbitrage-free and consistent across different leverage ratios. We propose a methodology for the pricing of options where the reference index is modeled by a Heston stochastic volatility model with jumps in the underlying. Our choice is motivated by the tractability of this model as well as its ability to reproduce, to a certain extent, observed features of the implied volatility surface. We start our analysis by considering the pricing problem when the reference index follows a Heston process without jumps. In this case, the LETF price also follows the Heston dynamics. This feature of *leverage-invariance* allows for tractability and efficient numerical pricing of options on LETFs. Thus, we analyze a practical calibration example under this framework and discuss our results. More specifically, we obtain the option prices for six LETFs on the reference S&P500 index and calibrate a Heston model on each of them separately. We then obtain the reference index model parameters implied by each of these calibrations. This allows us to compare their respective implied volatility surfaces against each other. We find that the model parameters agree to a significant extent, with the exception of a non-negligible difference between long and short LETFs. To facilitate the price comparison, we employ the concept of moneyness scaling and illustrate how to match the implied volatility skews for LETFs with different leverage ratios.

We further incorporate random jumps in the dynamics of the reference index. In contrast to the basic Heston model, the LETF dynamics are no longer of the same kind. While the continuous part of the dynamics is still Heston, the jump part is generally not of the same kind as that of the reference index. As a consequence, the model loses some of its tractability and a new pricing algorithm is required. We propose a numerical transform method and show that it is possible to price the options with the same computational complexity as in the Heston case, as well as for general distributions of the jump size. To prove the validity of our results, we then analyze a pricing example and compare it with results from the literature. We find that our algorithm performs well and with improved accuracy compared to existing methods. Since LETF providers may buy insurance against adverse moves in the underlier (see Section 4.7 for a detailed discussion and references), the protection by capping the LETF gains during favorable moves of the reference index. We thus incorporate the effects of such caps and floors on the LETF returns dynamics. We are able to include these effects in our pricing algorithm and evaluate their effects on the pricing of LETF options.

Chapter 2
Price Dynamics of Leveraged ETFs

In this chapter, we investigate the empirical returns of LETFs and present models for their price dynamics. We highlight the effects of leverage ratios and holding horizon on returns. An series of empirical studies are conducted to examine the tracking performance of LETFs and evaluate various leverage replication strategies. The unique characteristics of LETF price evolution motivate us to construct and backtest static delta-neutral long-volatility strategies for LETF portfolios.

2.1 Returns of Leveraged ETFs

To examine the returns of leveraged ETFs, let us consider the following illustrative example. By design, an LETF seeks to provide a constant multiple of the daily returns of an underlying index or asset. Let β be the leverage ratio stated by the LETF, and R_j the daily return of the reference. Ideally, the LETF value on day n, denoted by L_n, is

$$L_n = L_0 \cdot \prod_{j=1}^{n} (1 + \beta R_j). \tag{2.1}$$

We call this the leveraged benchmark, and examine the empirical performance of various LETFs with respect to this benchmark.

For many investors, one appeal of LETFs is that leverage can amplify returns when the underlying is moving in the desired direction. Mathematically,

© The Author(s) 2016
T. Leung, M. Santoli, *Leveraged Exchange-Traded Funds*, SpringerBriefs in Quantitative Finance, DOI 10.1007/978-3-319-29094-2_2

we can see this as follows. Rearranging (2.1) and taking the derivative of the logarithm, we have

$$\frac{d}{d\beta}\left(\log\left(\frac{L_n}{L_0}\right)\right) = \sum_{j=1}^{n} \frac{R_j}{1 + \beta R_j}. \tag{2.2}$$

With a positive leverage ratio $\beta > 0$, if $R_j > 0$ for all j, then $\log\left(\frac{L_n}{L_0}\right)$, or equivalently the value L_n, is increasing in β. In other words, when the reference asset is increasing in value, a larger, positive leverage ratio is preferred. On the other hand, if $R_j < 0$ for all j, and $\beta < 0$, a more negative β increases $\log\left(\frac{L_n}{L_0}\right)$ and thus L_n. This means that when the reference asset is decreasing in value, a more negative leverage ratio yields a higher return.

The example below illustrates the consequences of maintaining a constant leverage in an environment with nondirectional movements:

Day	ETF	%-change	+2x LETF	%-change	−2x LETF	%-change
0	100		100		100	
1	98	-2%	96	-4%	104	4%
2	99.96	2%	99.84	4%	99.84	-4%
3	97.96	-2%	95.85	-4%	103.83	4%
4	99.92	2%	99.68	4%	99.68	-4%
5	97.92	-2%	95.69	-4%	103.67	4%
6	99.88	2%	99.52	4%	99.52	-4%

Even though the ETF records a tiny loss of 0.12% after 6 days, the +2x LETF ends up with a loss of 0.48%, which is greater (in absolute value) than 2 times the return (−0.12%) of the ETF. We can see this to be the case on any day (e.g., not just the terminal date) except for day 1. For example, on day 3, the ETF has a net loss of 2.04% and the LETF has a net loss of 4.15%, which is greater (in absolute value) than 4.08% (twice the absolute value of the return of the ETF). Furthermore, it might be intuitive that the −2x LETF should have a positive return when the ETF and LETF have negative returns, this is not true. At the terminal date, both the long and short LETFs have recorded net losses of 0.48%. Again, this occurs throughout the period as well, not just the terminal date. In addition to day 6, both the long and short LETFs as well as the ETF itself are in the black. These results are consequences of volatility decay.

Although long and short LETFs are expected to move in opposite directions daily by design, it is often possible for both LETFs to have negative cumulative returns when held over a longer horizon. Figure 2.1 shows the historical cumulative returns of the ProShares leveraged gold ETFs, UGL (+2x) and GLL (−2x), which seek the corresponding multiples of the daily performance of the gold spot, over the period from July 2013 to July 2014. From trading day 124 (1/24/2014) onward, GLL has a negative cumulative return. There are points after trading date 124 where UGL also has a negative cumulative return. In fact, it starts in the black on this date and continues to have a net loss until trading date 146 (2/12/2014). This occurs again a few times, another long stretch where both have a net loss is trading date 210 (5/15/2014) through 233 (6/18/2014).This observation, though maybe counter-intuitive at first glance, is a consequence of daily replication of leveraged returns. The value erosion tends to accelerate during periods of nondirectional movements.

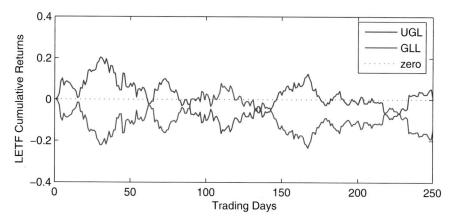

Fig. 2.1: UGL (+2x) and GLL (−2x) cumulative returns from July 2013 to July 2014. Observe that both UGL and GLL can give negative returns (below the dotted line of 0%) simultaneously over several periods in time.

A number of market observations suggest that LETFs exhibit value erosion as holding period increases, and it is more severe for highly leveraged ETFs. Let us illustrate by look at the empirical performance of LETFs over different holding horizons. Specifically, we compare the empirical returns of several major equity LETFs based on the S&P 500 index against multiples of the non-leveraged SPDR S&P 500 ETF (SPY).

In Figure 2.2, we present the returns of the ProShares Ultra S&P 500 ETF (SSO) with double long leverage ($\beta = +2$) and the ProShares Ultra-Short S&P 500 ETF (SDS) with double short leverage ($\beta = -2$), against ± 2 multiples of the SPY returns. We consider 1-day, 14-day, and 60-day rolling periods from September 29, 2010 to September 30, 2012. We observe from Figures 2.2(a)–2.2(b) that the returns fall along the straight line of slope 1. This reflects that both SSO and SDS are able to replicate, on a daily basis, the advertised multiple of the underlying ETF returns.

However, as the holding period lengthens to 14 days and 60 days, return discrepancies start to build (see Figures 2.2(c)–2.2(f)). In these cases, the LETF performance is often inferior to that of the underlying ETF, though the opposite could also happen, typically in a period with strong momentum. In general, a longer horizon also accumulates the erosion due to volatility drag. We shall investigate this more closely in subsequent sections.

In Figure 2.3, we present the same analysis between SPY and the triple-leveraged ETFs, namely, UPRO and SPXU, with leverage ratios $\beta = +3$ and -3, respectively. As we can see, the one-day returns are matched very closely, but longer horizons again lead to higher discrepancies in returns between the triple LETFs and the underlying. Comparing across leverage ratios, the underperformance over a 60-day period is more pronounced for the triple than the double leverage ratios (see Figures 2.2(e)–2.2(f) and 2.3(e)–2.3(f)). Furthermore, short LETFs tend to fail to replicate the required returns more often than their long leveraged counterparts.

2.2 Continuous-Time Model for Leveraged ETFs

We model the evolution of the reference index $(S_t)_{t \geq 0}$ by the stochastic differential equation:

$$dS_t = S_t \left(\mu_t \, dt + \sigma_t \, dW_t \right), \tag{2.3}$$

where $(W_t)_{t \geq 0}$ is a standard Brownian motion under the historical measure $\mathbb{P}$. The stochastic drift $(\mu_t)_{t \geq 0}$ represents the ex-dividend annualized growth rate process, and $(\sigma_t)_{t \geq 0}$ is the stochastic volatility process. At this point, we do not specify a parametric stochastic volatility model, though many well-known models, such as the Heston model as well as other stochastic or local volatility models, also fit within the above framework.

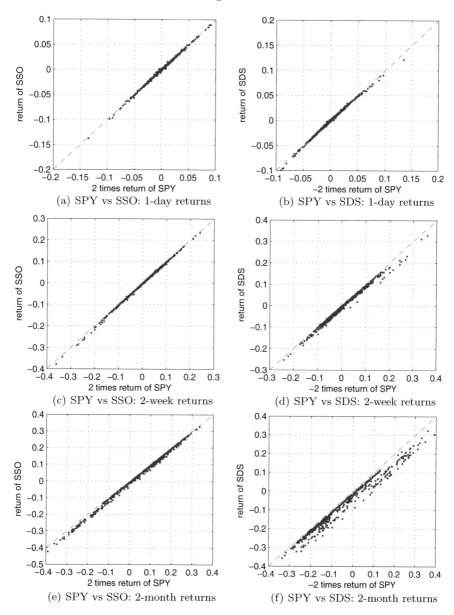

Fig. 2.2: 1-day (top), 2-week (center), and 2-month (bottom) returns of SPY against SSO (left) and SDS (right), in logarithmic scale. We considered 1-day, 2-week, and 2-month rolling periods from September 29, 2010 to September 30, 2012.

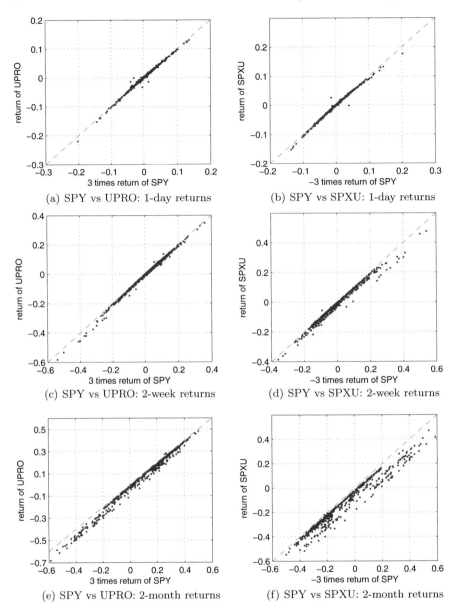

Fig. 2.3: 1-day (top), 2-week (center), and 2-month (bottom) returns of
SPY against UPRO (left) and SPXU (right), in logarithmic scale. We considered 1-day, 2-week, and 2-month rolling periods from September 29, 2010
to September 30, 2012.

Based on the reference index S, a long leveraged ETF $(L_t)_{t\geq 0}$ with leverage ratio $\beta \geq 1$ is constructed by simultaneously investing the amount βL_t (β times the fund value) in the underlying S, and borrowing the amount $(\beta-1)L_t$ at the interest rate $r \geq 0$. This is essentially a constant proportion trading strategy. As is typical for all ETFs, a small expense rate $f \geq 0$ is incurred. As a result, the β-LETF value evolves according to

$$dL_t = L_t \beta \frac{dS_t}{S_t} - L_t((\beta-1)r + f)\, dt. \tag{2.4}$$

On the other hand, a leveraged fund with a negative leverage ratio $\beta \leq -1$ involves taking a short position of amount $|\beta L_t|$ in S and keeping $(1-\beta)L_t$ in the money market account. The fund value $(L_t)_{t\geq 0}$ also satisfies (2.4) with $\beta \leq -1$. For some short LETFs, it would be appropriate to incorporate the rate of borrowing $\lambda \geq 0$ for short selling S. This can be achieved by replacing μ_t with $\mu_t + \lambda$ in (2.4) with $\beta \leq -1$. Theoretically, one can also construct constant proportion portfolio with $\beta \in (-1,1)$, but we do not discuss them since the most typical leverage ratios in practice are $\beta = 2,3$ (long) and $-2,-3$ (short).

For both long and short LETFs, we recognize from (2.4) that L can be expressed in terms of the underlying index S:

$$L_t = L_0 \exp\left(\int_0^t \left(\beta\mu_u - (\beta-1)r - f - \frac{\beta^2 \sigma_u^2}{2}\right) du + \int_0^t \beta\sigma_u dW_u \right) \tag{2.5}$$

$$= L_0 \left(\frac{S_t}{S_0}\right)^\beta \exp\left(-((\beta-1)r + f)t + \frac{1}{2}\beta(1-\beta)\int_0^t \sigma_u^2 du \right). \tag{2.6}$$

Taking (natural) log on both sides, we express the log-return of L in terms of that of S, namely,

$$\log\left(\frac{L_t}{L_0}\right) = \beta \log\left(\frac{S_t}{S_0}\right) - ((\beta-1)r + f)t + \frac{1}{2}\beta(1-\beta)\int_0^t \sigma_u^2 du. \tag{2.7}$$

In view of the second term, the long and short LETFs possess asymmetric return characteristics and volatility exposure. First, we observe that

$$\frac{1}{2}\beta(1-\beta) < 0, \quad \text{for} \quad \beta \notin [0,1],$$

so there is an erosion in log return proportional to the realized variance $\int_0^t \sigma_u^2 du$. Note that this effect is larger for a short LETF than its long leverage counterpart with the same magnitude. Since the realized variance is increasing in t, the value erosion, called *volatility decay*, is more significant over a longer holding horizon. Certainly, the expense fee also leads to decay in return. For more details on the derivation of (2.7) and its discrete-time analogue, we refer the reader to Avellaneda and Zhang (2010).

2.3 Empirical Leverage Ratio Estimation

Thus far the leverage ratio β has been taken as given. Indeed, it is advertised by the ETF provider as the target leverage ratio that the ETF seeks to achieve. In this section, we introduce a novel method to estimate the empirical leverage ratio realized by any given LETF.

In discrete time, with $\Delta t = 252^{-1}$, the k-day log-return of the LETF is given by

$$\log \frac{L_{t+k\Delta t}}{L_t} = \beta \log \frac{S_{t+k\Delta t}}{S_t} + \theta V_t^{(k)} + ((1-\beta)r - f)k\Delta t, \qquad (2.8)$$

where the realized variance is computed by

$$V_t^{(k)} = \sum_{i=0}^{k-1} (R_{t+i\Delta t}^S - \bar{R}_t^S)^2, \quad \text{with} \quad \bar{R}_t^S = \frac{1}{n} \sum_{i=0}^{n-1} R_{t+i\Delta t}^S,$$

and R_t^S is the daily return of the reference index at time t.

The log-return equation (2.7), or its discretized version, immediately suggests a linear regression:

$$\log \frac{L_t}{L_0} = \hat{\beta} \log \frac{S_t}{S_0} + \hat{\theta} V_t + \hat{c} + \epsilon, \qquad (2.9)$$

where $V_t = \int_0^t \sigma_u^2 du$ and $\epsilon \sim N(0,1)$. Therefore, this results in a linear model, from which one can estimate from historical LETF and reference prices the constant coefficients $\hat{\beta}$, $\hat{\theta}$, and $\hat{c}$.

However, there is a significant collinearity issue due to the strong dependence of the two variables $\log \frac{S_t}{S_0}$ and V_t. Therefore, the coefficients estimated using this standard approach are not reliable. Guo and Leung (2015) have conducted regressions for 22 commodity LETFs and illustrated the issue of collinearity. For the purpose of estimating β, this approach also leads to another problem. Indeed, the first coefficient in (2.9) represents the estimated leverage ratio $\hat{\beta}$, but the coefficient $\hat{\theta}$ also has a theoretical value in terms of leverage ratio, i.e., $\beta(1-\beta)/2$. Thus, one can back out another estimated leverage ratio $\tilde{\beta}$ from $\hat{\theta} = \tilde{\beta}(1-\tilde{\beta})/2$, but there is no guarantee that the resulting leverage ratio will equal $\hat{\beta}$. In fact, as Guo and Leung (2015) have shown for many LETFs, the two estimated leverage ratios are most certainly different and deviate significantly from the theoretical leverage ratio β. This leads to an important dilemma - which estimate should we use? Is either estimate reliable?

Motivated by the above observations, we now discuss a new way to determine the realized leverage ratio. One of our main objectives is to give a single optimized estimate. To this end, we seek to find the leverage ratio β that minimizes the sum of squared differences between the realized LETF log-returns and the theoretical LETF log-return based on (2.7). As such, we solve the optimization problem:

$$\min_{\beta \in \mathbb{R}} \sum_{i=1}^{n} (y_i - f_i(\beta))^2$$

where $(y_i)_{i=1,\ldots,n}$ are the empirical log-returns of the LETF, and $(f_i(\beta))_{i=1,\ldots,n}$ are the theoretical returns given by

$$f_i(\beta) = \beta x_i - \frac{\beta(\beta-1)}{2} v_i + ((1-\beta)r - f)\Delta T \tag{2.10}$$

$$= \beta(x_i - r\Delta T) - \frac{\beta(\beta-1)}{2} v_i + (r - f)\Delta T, \tag{2.11}$$

where each $f_i(\beta)$ requires the log-return of the reference x_i, and the realized variance v_i over the same period of length ΔT (see (2.7)).

The optimal leverage ratio is found from the first-order optimality condition:

$$\sum_{i=1}^{n} (y_i - f_i(\beta))(x_i - r\Delta T - \beta v_i + \frac{1}{2} v_i) = 0.$$

We expand the left-hand side to get

$$\sum_{i=1}^{n}(y_i - f_i(\beta))(x_i - r\Delta T - \beta v_i + \frac{1}{2}v_i)$$

$$=\sum_{i=1}^{n}(y_i - \beta(x_i - r\Delta T) + \frac{\beta^2}{2}v_i - \frac{\beta}{2}v_i - (r - f)\Delta T)(x_i - r\Delta T - \beta v_i + \frac{1}{2}v_i)$$

$$=\sum_{i=1}^{n}(y_i - (r - f)\Delta T - \beta(x_i - r\Delta T + \frac{v_i}{2}) + \frac{\beta^2}{2}v_i)(x_i - r\Delta T - \beta v_i + \frac{1}{2}v_i)$$

$$=\left(-\sum_{i=1}^{n}\frac{v_i^2}{2}\right)\beta^3 + \left(\sum_{i=1}^{n}\frac{3}{2}(x_i - r\Delta T)v_i + v_i^2\right)\beta^2$$

$$+\left(\sum_{i=1}^{n}-((x_i - r\Delta T) + \frac{1}{2}v_i)^2 + v_i((r - f)\Delta T - y_i)\right)\beta$$

$$+\left(\sum_{i=1}^{n}(y_i - (r - f)\Delta T)((x_i - r\Delta T) + \frac{1}{2}v_i)\right).$$

As a result, the optimality condition reduces to the cubic equation

$$A\beta^3 + B\beta^2 + C\beta + D = 0, \tag{2.12}$$

where the constant coefficients are given by

$$A = -\sum_{i=1}^{n}\frac{v_i^2}{2},$$

$$B = \sum_{i=1}^{n}\frac{3}{2}(x_i - r\Delta T)v_i + v_i^2,$$

$$C = \sum_{i=1}^{n}-((x_i - r\Delta T) + \frac{1}{2}v_i)^2 + v_i((r - f)\Delta T - y_i,$$

$$D = \sum_{i=1}^{n}(y_i - (r - f)\Delta T)((x_i - r\Delta T) + \frac{1}{2}v_i).$$

Dividing by A in (2.12), we find the root of the equation:

$$\beta^3 + b\beta^2 + c\beta + d = 0,$$

with the obvious definitions for b, c, and d here. The discriminant of this cubic polynomial is

$$\Delta = 18bcd - 4b^3d + b^2c^2 - 4c^3 - 27d^3.$$

which tells us one of the following three cases:

1. If $\Delta > 0$, then the equation has 3 distinct real roots.
2. If $\Delta = 0$, then the equation has a multiple root and all its roots are real.
3. If $\Delta < 0$, then the equation has one real root and two complex conjugate roots.

By the well-known Cardano's method for cubic polynomials, the explicit solutions when $\Delta < 0$ are given by

$$\beta_1 = u_0 + u_1 - \frac{b}{3}, \quad \beta_{2,3} = -\frac{1}{2}(u_0 + u_1) \pm \frac{i\sqrt{3}}{2}(u_0 - u_1) - \frac{b}{3},$$

where

$$u_i = \sqrt[3]{-\frac{p}{2} + (-1)^i \sqrt{\frac{p^2}{4} + \frac{q^3}{27}}}, \quad i = 0, 1,$$

$$p = \frac{2b^3 - 9bc + 27d}{27}, \quad q = \frac{3c - b^2}{3}.$$

If $\Delta = 0$ (iff $\frac{p^2}{4} + \frac{q^3}{27} = 0$), then the roots are real and at least two are the same:

$$-2\sqrt{-\frac{q}{3}} - \frac{b}{3}, \quad \sqrt{-\frac{q}{3}} - \frac{b}{3}, \quad \sqrt{-\frac{q}{3}} - \frac{b}{3} \quad \text{if } p > 0,$$

$$2\sqrt{-\frac{q}{3}} - \frac{b}{3}, \quad -\sqrt{-\frac{q}{3}} - \frac{b}{3}, \quad -\sqrt{-\frac{q}{3}} - \frac{b}{3} \quad \text{if } p < 0,$$

$$0, \quad 0, \quad 0 \quad \text{if } p = 0.$$

If $\Delta > 0$ (iff $\frac{p^2}{4} + \frac{q^3}{27} < 0$), then the roots are real and can be expressed as

$$\beta_n = 2\sqrt{-\frac{q}{3}} \cos\left(\frac{\gamma}{3} + \frac{2n\pi}{3}\right), \quad n = 0, 1, 2$$

where

$$\gamma = \cos^{-1} \sqrt{\frac{p^2/4}{-q^3/27}}.$$

Alternatively, one can obtain numerical solutions using the root finding methods in a commercial computational software.

In Table 2.1, we apply our estimation method to six S&P500 based LETFs using the returns observed during 1/1/2013 to 5/31/2015. We can see that the leverage ratio β_{cub} estimated from our optimization method is extremely close to the target multiple β. The leverage ratio β_{reg} estimated from linear regression is also close to β. However, the associated realized volatility coefficient θ_{reg} from regression is not close to its theoretical value $\theta = (\beta - \beta^2)/2$, especially for the three LETFs: SSO, UPRO, and SH, where the absolute percentage errors are 69%, 14%, and 46%, respectively.

LETFs	β	θ	β_{cub}	β_{reg}	θ_{cub}	θ_{reg}
SPY	1	0	1.0004	1.0009	−0.0001	−0.1091
SSO	2	−1	2.0001	1.9910	−1.0001	−1.6918
UPRO	3	−3	3.0110	3.0003	−3.0275	−3.4178
SH	−1	−1	−1.0012	−0.9920	−1.0018	−0.5439
SDS	−2	−3	−2.0062	−1.9906	−3.0156	−2.9451
SPXU	−3	−6	−2.9965	−2.9708	−5.9879	−5.9619

Table 2.1: Estimated parameters using cubic root-finding ($\beta_{cub}, \theta_{cub}$) and linear regression ($\beta_{reg}, \theta_{reg}$) for six S&P500 based LETFs. Recall that theoretically $\theta = (\beta - \beta^2)/2$. Returns are computed based on 5-day holding periods during 01/01/2013 to 05/31/2015.

The leverage ratio estimation requires not only the returns of the LETFs but also the empirical variance of the reference index. Therefore, we need to partition the entire period into subintervals of n-days and compute the realized variance for each n-day horizon. If the sample period is short, e.g., a year or less, then this will yield a small number of data points and cause problems for the linear regression approach. This issue may arise for many LETFs that were introduced to the market only recently. Even for LETFs with longer histories, it would be useful to compare the estimated leverage ratio over different periods, for example, quarter by quarter. Again, this implies partitioning into short periods with low number of data points.

A major strength of our method, compared to linear regression, is that we do not need to work with a long sample period. This is mainly because our method involves finding a single variable β_{cub} by minimizing a univariate quadratic function. Once the estimated leverage ratio β_{cub} is obtained, the other coefficient θ_{cub} is instantly computed by $\theta_{cub} = (\beta_{cub} - \beta_{cub}^2)/2$ and is thus guaranteed to be consistent. In contrast, the linear regression involves finding the optimal pair $(\beta_{reg}, \theta_{reg})$ simultaneously from data, without constraining them to satisfy the known relationship.

Table 2.2 summarizes the estimated leverage ratios for six S&P500 LETFs from four quarters in 2014. With just over 60 trading days in each quarter, the linear regression method fails to return accurate or stable estimates for the realized variance coefficient θ_{reg}. In contrast, the cubic root-finding method generates a series of stable θ_{cub} that are very close to the theoretical value, over the quarters for each of the six LETFs.

Leveraged ETFs are commonly advertised to generate a prespecified multiple of the reference index return on a daily basis, regardless of the movements of the reference or market conditions. As is well known, the accuracy of the promised return replication varies across leveraged ETFs. Even if we focus on a single leveraged ETF, its deviation from the stated objective may change over time.

We want to measure the empirical leverage ratio conditioned on the *sign* of the returns of the reference index. To this end, we apply our cubic root-finding method for the periods during which the reference index experiences positive returns, and separately, for the periods with negative reference returns.

The results are quite surprising. Figure 2.4(a) displays the empirical leverage ratios estimated from 1/1/2013 to 5/31/2015, with different sampling subintervals (in days). When the reference returns are positive, the leverage ratio tends to be higher than the stated multiple (+2) for the double-long leveraged ETF, SSO. In contrast, this LETF realizes a leverage ratio less than +2 when the reference returns are negative.

If we turn to Figure 2.5(a) for the double-short leveraged ETF, SDS, we observe that the deviation from the stated leverage ratio −2 is more significant. When the reference returns are positive (resp. negative), the leverage ratio tends to be more (resp. less) negative than −2. This can be interpreted as the two LETFs tend to *over-leverage* when the reference index

LETFs	Qtr	β_{cub}	β_{reg}	θ_{cub}	θ_{reg}
SPY	1	0.9956	0.9938	0.0022	−0.4319
($\beta = +1$)	2	0.9924	1.0102	0.0038	2.3251
($\theta = 0$)	3	1.0080	1.0079	−0.0040	−0.0163
	4	0.9670	0.9634	0.0159	−1.1210
SSO	1	1.9941	1.9761	−0.9911	−4.4384
($\beta = +2$)	2	1.9999	1.9599	−0.9998	−5.1184
($\theta = -1$)	3	1.9947	1.9774	−0.9921	−4.1630
	4	1.9818	1.9975	−0.9729	−1.6769
UPRO	1	2.9801	2.9583	−2.9505	−7.0032
($\beta = +3$)	2	3.0236	2.9723	−3.0594	−8.1769
($\theta = -3$)	3	3.0047	2.9850	−3.0118	−6.7569
	4	2.9801	2.9728	−2.9504	−3.5249
SH	1	−0.9927	−0.9914	−0.9891	−1.3642
($\beta = -1$)	2	−1.0166	−1.0110	−1.0251	−1.3263
($\theta = -1$)	3	−0.9906	−0.9852	−0.9859	−0.0514
	4	−1.0060	−1.0025	−1.0090	−1.7533
SDS	1	−1.9821	−1.9741	−2.9555	−2.5680
($\beta = -2$)	2	−2.0148	−1.9757	−3.0380	−0.1355
($\theta = -3$)	3	−1.9775	−1.9711	−2.9440	−1.9718
	4	−2.0113	−2.0009	−3.0284	−3.2447
SPXU	1	−2.9693	−2.9497	−5.8932	−3.6919
($\beta = -3$)	2	−3.0198	−2.9583	−6.0697	−1.3943
($\theta = -6$)	3	−2.9601	−2.9384	−5.8610	−2.3462
	4	−3.0161	−2.9974	−6.0564	−6.2966

Table 2.2: Estimated parameters using cubic root-finding ($\beta_{cub}, \theta_{cub}$) and linear regression ($\beta_{reg}, \theta_{reg}$) for six S&P500 based LETFs. Recall that theoretically $\theta = (\beta - \beta^2)/2$. Returns are computed based on 3-day holding periods over the 4 quarters in 2014.

experiences positive returns and *under-leverage* when the reference loses value. Moreover, if we compare our results to those using linear regression in Figures 2.4(b) and 2.5(b), the latter approach fails to generate a clear pattern. This illustrates another useful application of our method and its advantage over linear regression.

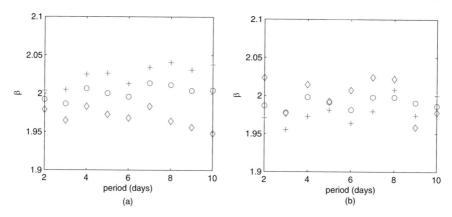

Fig. 2.4: Empirical leverage ratios for the double-long ($\beta = +2$) LETF, SSO, estimated using (a) the cubic root-finding approach, and (b) linear regression. Returns during $01/01/13$ to $05/31/15$ are used.

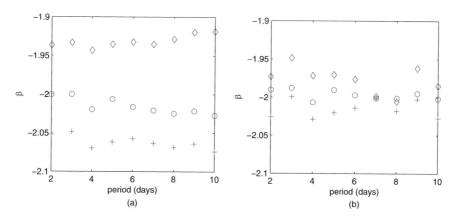

Fig. 2.5: Empirical leverage ratios for the double-short ($\beta = -2$) LETF, SDS, estimated using (a) the cubic root-finding approach, and (b) linear regression. Returns during $01/01/13$ to $05/31/15$ are used.

2.4 Dynamic Leveraged Futures Portfolio

For some ETFs, the reference asset or index may be very illiquid or not even traded. This can be a reason that the leveraged ETF price evolution may deviate from the dynamics of the leveraged portfolio described in (2.4). For LETFs that track commodity spot prices, one can alternatively trade the commodity futures to generate the required leverage.

In this section we analyze the returns and tracking performances of various leveraged ETFs. From historical prices of each LETF, we conduct an estimation of the leverage ratio and investigate the potential deviation from the target leverage ratio. Moreover, we construct a number of static portfolios with futures contracts to seek replication of some leveraged benchmarks. However, the static portfolios fail to effectively track the leveraged benchmarks. This motivates us to consider a dynamic portfolio with futures, which turns out to have a much better tracking performance.

First, we conduct a regression analysis and the results are given in Table 2.3. Each slope is approximately equal to the LETF's target leverage ratio. In principle, if each (L)ETF is able to generate the desired multiple of daily returns, the slopes of the regression should be equal to the various leverage ratios. In this table, we give an additional two columns for the t-statistic and p-value for testing the hypothesis: $\{H_0 : \text{slope} = \beta\}$ vs. $\{H_1 : \text{slope} \neq \beta\}$. Here, β is the target leverage ratio. We can see that each p-value is larger than 0.05 and therefore conclude that statistically, each (L)ETF does not differ from its target leverage ratio. This demonstrates us that the (L)ETFs are performing exactly as desired, at least on a daily basis.

(L)ETF	Slope	Intercept	t-stat	p-value	R^2	$RMSE$
GLD	1.0054	$-1.64 \cdot 10^{-5}$	1.4269	0.1538	0.9806	0.0016
UGL	2.0057	$-1.31 \cdot 10^{-4}$	0.7332	0.4636	0.9793	0.0034
GLL	-2.0056	$-1.06 \cdot 10^{-4}$	0.6709	0.5024	0.9767	0.0036
UGLD	2.9936	$-1.99 \cdot 10^{-4}$	0.4521	0.6513	0.9848	0.0042
DGLD	-2.9753	$-1.69 \cdot 10^{-5}$	0.9744	0.3302	0.9526	0.0075

Table 2.3: A summary of the regression coefficients and measures of goodness of fit for regressing one-day returns of (L)ETFs versus spot gold. We include 2 additional columns for the t-statistic and p-value for testing the hypothesis that the slope equals the leverage ratio in each case.

We see also that the R^2 values for each regression are quite high, all above 95%. Next, we compare the long and short LETFs for a fixed $|\beta| \in \{2, 3\}$. The short LETF tends to have a higher RMSE and lower R^2 value. Finally, we see in general that as the leverage ratio increases in absolute value, there is a higher RMSE. One possible explanation is that the benchmark is leveraged, and this could magnify the tracking error.

Let us analyze the effects of changing the holding period. In Table 2.4 we give the slopes and intercepts for the regressions of each (L)ETF's return versus the spot return while varying the holding period between 1 and 5 days. Our computations show the R^2 values are all above 95%. We can see that all the slopes are approximately equal to the target leverage ratio of the (L)ETF. However we notice that in general the intercepts get more negative as the holding period is lengthened. Although they are still quite small, they become more significant as the holding period increases. Our calculations show that the p-values for testing the hypothesis: $\{H_0 : \text{intercept} = 0\}$ vs. $\{H_1 : \text{intercept} \neq 0\}$ generally tend to decrease for each (L)ETF. In fact, for UGL the intercepts turn out to be statistically different from 0 (at the 5% level) for holding periods of 3, 4, and 5 days with p-values of 1.37%, 0.57%, and 0.33%, respectively. This is consistent with the volatility decay discussed above. We saw there an example where over shorter periods, the LETF tracks its leverage ratio well, but over a longer period it tends to lose money when there is high volatility. The intercepts being different from 0 is akin to the volatility decay in the following sense. Over longer periods, the regressions show that we require more information than just the gold return to predict the LETF return.

To compare the performance of the LETF versus the target multiple of the spot return, we also report in Table 2.4 the average return differential defined by

$$\overline{RD} = \frac{1}{m} \sum_{j=1}^{m} \left(R_j^{(L)} - \beta \cdot R_j^{(G)} \right), \tag{2.13}$$

where m is the number of the periods, $R_j^{(L)}$ is the LETF's return over the holding period, and $R_j^{(G)}$ is the spot's return over the holding period. We find this to be increasing (in absolute value) with the holding period length. That is, as we hold the LETF longer, it tends to increasingly underperform with respect to the multiple of the underlying return, on average. This is exactly the same notion described above, since over time, the volatility of the underlying causes the LETF to erode in value.

	Days	UGL	GLL	UGLD	DGLD	GLD
Slope	1	2.0057	-2.0056	2.9916	-2.9636	1.0054
	2	2.0083	-2.0040	2.9202	-3.0243	1.0048
	3	1.9777	-1.9966	2.9452	-3.0669	0.9919
	4	2.0007	-2.0091	2.9785	-3.0431	1.0021
	5	2.0208	-2.0327	2.8948	-3.0727	1.0104
Intercept $(\cdot 10^{-4})$	1	-1.3107	-1.0551	-1.9861	-0.1691	-0.1643
	2	-2.7328	-2.2881	-4.2312	-1.8155	-0.3339
	3	-4.0642	-3.9663	-4.9799	-1.0563	-0.4084
	4	-5.4443	-4.6217	-8.0848	-2.6997	-0.6613
	5	-6.8161	-4.8281	-8.0342	-0.3077	-0.9236
$\overline{RD}\,(\cdot 10^{-3})$	1	-0.1289	-0.1076	-0.1967	-0.0241	-0.0144
	2	-0.2668	-0.2319	-0.4400	-0.1908	-0.0296
	3	-0.4325	-0.3927	-0.5351	-0.0846	-0.0503
	4	-0.5433	-0.4765	-0.8169	-0.2785	-0.0629
	5	-0.6408	-0.5471	-0.7918	0.0717	-0.0720

Table 2.4: A summary of the slopes and intercepts from the regressions of LETF returns versus gold returns, as well as the average return differential ($\overline{RD}$) over different holding periods.

2.4.1 Static Leverage Replication

We consider the problem of static leverage replication and seek an optimal static portfolio of futures which minimizes SSE. Let k be the number of futures contracts and $\mathbf{w} := (w_0, \ldots, w_k)$ be the real-valued vector of portfolio weights. As before, w_0 represents the weight given to the money market account. We seek the weights which minimize SSE over the 5-year period 12/22/2008 through 12/22/2013. Thus, we are led to the same constrained least squares optimization problem:

$$\min_{\mathbf{w}\in\mathbb{R}^{k+1}} \|\mathbf{Cw} - \mathbf{L}\|^2$$
$$\text{s.t.} \quad \sum_{j=0}^{k} w_j = 1 \tag{2.14}$$

Again, the matrix $\mathbf{C}$ contains as columns, the historical prices of the various futures contracts and the money market account. Here, the vector $\mathbf{L}$ contains the historical prices of the leveraged benchmark in (2.1). Without loss of generality, we normalize the prices by \$1000 so that our solution will give us a set of weights on each instrument.

To compare the tracking error of our optimized portfolios to that of investments in the LETFs, we will perform an out of sample analysis over the period 12/23/2013 through 7/14/2014 and see how \$1000 invested in the LETFs and \$1000 invested in our optimal portfolios perform. In order to quantify the performance we use the same root mean square error

$$RMSE = \sqrt{\frac{1}{n}\sum_{j=1}^{n}(V_j - L_j)^2}, \qquad (2.15)$$

where V_j is the dollar value of the portfolio on trading day j, while L_j is the dollar value of the leveraged benchmark on trading day j. Now, we present the results for the optimization and in sample/out of sample RMSE.

UGL(+2x)	Futures	w_0	w_1	w_2	$RMSE$ (in)	$RMSE$ (out)
1 Futures	1-m	-1.4816	2.4816	-	153.2015	41.4853
	2-m	-1.5760	2.5760	-	140.4125	49.5663
	6-m	-1.5811	2.5811	-	138.6312	47.4595
	12-m	-1.6263	2.6263	-	134.5068	48.2985
2 Futures	1-m, 2-m	-1.9450	-9.8944	12.8394	114.3095	81.3697
	1-m, 6-m	-1.9118	-8.4345	11.3463	113.6758	67.3943
	1-m, 12-m	-2.0222	-6.9279	9.9501	108.2990	67.1578
	2-m, 6-m	-1.6454	-34.2620	36.9073	126.6207	22.6495
	2-m, 12-m	-1.9910	-18.9915	21.9825	111.7505	40.9285
	6-m, 12-m	-2.2434	-35.6692	38.9126	102.8628	62.5383

Table 2.5: A summary of the weights and in/out of sample RMSEs for portfolios of 1 and 2 futures contracts which attempt to replicate a leveraged benchmark with $\beta = 2$. By comparison, the +2x LETF, UGL has an out of sample RMSE of only 5.5249.

GLL(−2x) Futures		w_0	w_1	w_2	$RMSE$ (in)	$RMSE$ (out)
1 Futures	1-m	1.9754	-0.9754	-	152.3335	76.1438
	2-m	2.0107	-1.0107	-	155.7615	73.1470
	6-m	2.0123	-1.0123	-	156.4393	74.0060
	12-m	2.0293	-1.0293	-	158.0679	73.7924
2 Futures	1-m, 2-m	1.6951	-8.4629	7.7678	139.2744	100.1009
	1-m, 6-m	1.7051	-7.8346	7.1295	137.9877	92.2112
	1-m, 12-m	1.6077	-7.3754	6.7677	133.3171	93.1121
	2-m, 6-m	1.9390	-39.0864	38.1474	142.5732	43.4071
	2-m, 12-m	1.5768	-23.5616	22.9848	127.9063	62.1042
	6-m, 12-m	1.2789	-43.3691	-43.0902	117.8184	88.3599

Table 2.6: A summary of the weights and in/out of sample RMSEs for portfolios of 1 and 2 futures contracts which attempt to replicate a leveraged benchmark with $\beta = -2$. By comparison, the −2x LETF, GLL has an out of sample RMSE of only 4.7627.

UGLD(+3x) Futures		w_0	w_1	w_2	$RMSE$ (in)	$RMSE$ (out)
1 Futures	1-m	-3.3370	4.3370	-	555.4915	111.5113
	2-m	-3.5063	4.5063	-	529.9086	125.9864
	6-m	-3.5156	4.5156	-	526.5845	122.3318
	12-m	-3.5959	4.5959	-	518.9690	123.8302
2 Futures	1-m, 2-m	-5.1815	-44.9250	51.1065	379.1168	270.5829
	1-m, 6-m	-5.0268	-38.5351	44.5619	381.9211	213.6217
	1-m, 12-m	-5.4197	-31.9108	38.3305	366.4976	210.9125
	2-m, 6-m	-3.7808	-141.3154	146.0961	472.3290	48.0413
	2-m, 12-m	-5.1256	-79.6610	85.7866	413.1968	98.3373
	6-m, 12-m	-6.1401	-147.0441	154.1842	376.4005	185.8717

Table 2.7: A summary of the weights and in/out of sample RMSEs for portfolios of 1 and 2 futures contracts which attempt to replicate a leveraged benchmark with $\beta = 3$. By comparison, the +3x LETF, UGLD has an out of sample RMSE of only 6.0813.

DGLD(-3x) Futures		w_0	w_1	w_2	$RMSE$ (in)	$RMSE$ (out)
1 Futures	1-m	2.1474	-1.1474	-	222.5908	135.6776
	2-m	2.1889	-1.1889	-	225.7671	132.1496
	6-m	2.1908	-1.1908	-	226.4291	133.1623
	12-m	2.2106	-1.2106	-	228.1371	132.9166
2 Futures	1-m, 2-m	1.8265	-9.7161	8.8896	211.0905	163.0473
	1-m, 6-m	1.8353	-9.0649	8.2296	209.7556	154.1627
	1-m, 12-m	1.7078	-8.7977	8.0899	204.4129	155.8254
	2-m, 6-m	2.1026	-46.9910	45.8884	212.7849	95.2484
	2-m, 12-m	1.6398	-29.7260	29.0862	195.7478	117.3542
	6-m, 12-m	1.2446	-55.8313	55.5867	183.4220	150.5826

Table 2.8: A summary of the weights and in/out of sample RMSEs for portfolios of 1 and 2 futures contracts which attempt to replicate a leveraged benchmark with $\beta = -3$. By comparison, the -3x LETF, DGLD has an out of sample RMSE of only 4.4372.

The static portfolios do not replicate the leveraged benchmark well here. In Tables 2.5, 2.6, 2.7, and 2.8, the RMSE values are quite large for all the portfolios. The minimum RMSE for any portfolio of futures trying to replicate any leveraged benchmark is 22.6495 (achieved by a portfolio of 2-month and 6-month futures attempting to replicate a $+2$x investment in gold) and by comparison the maximum RMSE for any LETF trying to replicate its respective leveraged investment is 6.0813. (This is achieved by UGLD, which tracks a $+3$x investment in gold.) Unlike the unleveraged investment, the money market account is extensively used throughout the various portfolios. This is interesting but also logical. Indeed, in order to create leverage, the portfolio must either borrow if $\beta > 0$ or invest in the money market account if $\beta < 0$.

Furthermore, the optimal weights tend to lead to over/under-leveraging. Since we are considering an investment in gold, the sum of the weights on the futures (which are instruments for investment in gold) can be interpreted as the leverage on the portfolio. Since all the weights sum to 1, we can compute the approximate leverage as $1 - w_0$. For the $+2$x and $+3$x investments, these values are all larger than 2 and 3, respectively. For the -2x and -3x investments, these values are all smaller (in absolute value) than -2 and -3, respectively. Thus we see that the long portfolios tend to be over-leveraged, while the short portfolios tend to be under-leveraged.

The optimization procedure has led to some rather uneven portfolio weights. For example, the optimal portfolio of 6-month and 12-month futures that attempts to replicate a +3x investment in spot gold requires the following transactions at inception: borrow \$6,140.11 from the money market account, short \$147,044.08 in 6-month futures and long \$154,184.19 in 12-month futures. In practice this would not be possible in the marketplace due to position limits that may be in place.

2.4.2 Dynamic Leverage Replication

To improve upon the replication in Section 2.4.1, we now consider an example in which we construct a dynamic portfolio with the front-month futures contract and cash. Let P_t be our portfolio value at time t. At every point in time, the portfolio invests β times the value of the fund in the futures contract in order to achieve the required leverage. As a result, the value of our portfolio is similar to (2.4) but the reference price S (gold spot in this example) is replaced by the front-month futures price.[1]

To quantify our portfolio's replicating ability we will use the same root mean squared error:

$$RMSE = \sqrt{\frac{1}{n}\sum_{j=1}^{n}(P_j - L_j)^2}, \qquad (2.16)$$

where L_j is the value of a leveraged investment in gold and P_j is the value of our portfolio, each at trading day j. For this dynamic portfolio, there is no sample from which we will need to draw our weights or train our model in any way. Therefore, we can look at any conceivable time period and compare how the LETF (L) or leveraged portfolio (P) performs using the metric in (2.16).

For this tracking metric, we consider the period 1/3/2012 (first trading day of 2012) to 7/14/2014 for gold leveraged ETFs listed in Table 2.9. The results are shown in Table 2.10. The portfolio RMSEs range between 0.687%

---

[1] Leung and Ward (2015) show that the front-month futures is empirically most effective in replicating the spot gold price.

LETF	Reference	Underlying	Issuer	β	Fee	Inception
GLD	GOLDLNPM	Gold Bullion	iShares	1	0.40%	11/18/2004
UGL	GOLDLNPM	Gold Bullion	ProShares	2	0.95%	12/01/2008
GLL	GOLDLNPM	Gold Bullion	ProShares	-2	0.95%	12/01/2008
UGLD	SPGSGCP	Gold Bullion	VelocityShares	3	1.35%	10/17/2011
DGLD	SPGSGCP	Gold Bullion	VelocityShares	-3	1.35%	10/17/2011

Table 2.9: A summary of the gold LETFs, along with the non-leveraged ETF (GLD). The LETFs with higher absolute leverage ratios, $|\beta| \in \{2, 3\}$, tend to have higher expense fees.

and 3.291%, which are smaller than LETF RMSEs which range between 1.87% and 4.338%. Overall, we see that the portfolio RMSEs are lower than the LETF RMSEs. Indeed, we see that our dynamic portfolio is able to track the target leveraged index quite well according to the RMSE values for $\beta \in \{2, -2, 3\}$. However the tracking is not as strong for $\beta = -3$. Nonetheless, the value is quite small and not that far off from the LETF RMSE.

In Figure 2.6, we see the time evolution for both the dynamic portfolio and GLL compared to the -2x benchmark. It is visible that the LETF tends to underperform the benchmark and the difference worsens over time. On the other hand, the portfolio tends to stay close to the benchmark over the entire period. Though not reported here, we observe similar patterns for other gold LETFs.

In Table 2.10, we also give the annual returns for each asset for the years 2011, 2012, and 2013. For UGLD and DGLD we do not have data for the full year of 2011 (its issue date was $10/17/2011$) so we do not have annual returns for these LETFs in 2011. The dynamic portfolio returns range between -69.22% and 107.54% while the LETF returns range between -69.90% and 106.16%. Comparing each year and leverage ratio pair, we find that, except for $\beta = -3$ in 2012, our portfolio outperforms the LETF in each year. Thus, we have shown that in general a dynamic portfolio consisting of just one futures contract can not only more closely track the target leveraged index, but it also outperforms the respective LETF.

β	Asset	RMSE	Annual Return (%)		
			2011	2012	2013
+2x	UGL	30.33	12.90	2.81	-52.31
	Portfolio	6.87	15.23	6.29	-51.83
−2x	GLL	40.12	-29.43	-16.40	67.82
	Portfolio	15.87	-27.06	-14.89	70.57
+3x	UGLD	43.38	-	0.41	-69.90
	Portfolio	12.55	-	5.29	-69.22
−3x	DGLD	18.70	-	-23.57	106.16
	Portfolio	32.91	-	-24.51	107.54

Table 2.10: A summary of the annual returns (over the periods: 1/3/2011 to 12/31/2011, 1/3/2012 to 12/31/2012, and 1/2/2013 to 12/31/2013) and RMSE for each LETF and a dynamic portfolio of front-month futures and cash. RMSE values are calculated over the period 1/3/2012 to 7/14/2014.

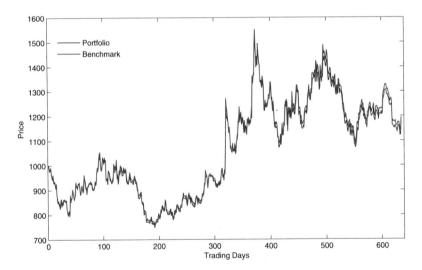

Fig. 2.6: Time evolution of our dynamic portfolio of front-month futures and cash (top) compared to the −2x benchmark and GLL (bottom) compared to the −2x benchmark. Time period displayed is 1/3/2012 to 7/14/2014.

2.5 Static Delta-Neutral Long-Volatility LETF Portfolios

The advent of ETFs has facilitated pairs trading in industry since many ETFs are designed to track identical or similar indexes and assets. For instance, Triantafyllopoulos and Montana (2009) study the mean-reverting dynamics of the spreads between commodity ETFs, and Leung and Li (2015b) derive the optimal timing strategies to trade an ETF pair.

LETFs can also be used in combination to construct various portfolios. Some strategies are designed accounting for the value erosion due to volatility decay associated with LETFs. In this section, we discuss how to construct LETF portfolios that are insensitive to the changes of the underlying (delta-neutral) but has a long volatility exposure.

For our analysis in this section, the reference index S is assumed to follow a general diffusion price dynamics described in (2.3), and the LETF value is given by

$$L_t = L_0 \left(\frac{S_t}{S_0} \right)^\beta \exp \left((-(\beta - 1)r - f)\,t - \frac{1}{2}\beta(\beta - 1)V_t \right), \qquad (2.17)$$

where $V_t = \int_0^t \sigma_u^2\, du$ is the realized variance of S up to time t.

Taking advantage of the volatility decay, a well-known trading strategy used by practitioners involves shorting a $\pm\beta$ pair of LETFs with the same reference, as discussed in Leung and Santoli (2012); Mackintosh and Lin (2010); Mason et al. (2010). Since the LETFs have opposite daily returns on the same reference index, the portfolio has little exposure to the reference index as long as the holding period is sufficiently short. With this strategy, the volatility decay can in fact help generate profit. However, the portfolio is exposed to risk during periods of low volatility and high trending even if we assume that the LETFs are tracking perfectly. We now describe an extension of this trading strategy by allowing the positive and negative leverage ratios to differ. As we determine the portfolio weights to eliminate the dependence on the reference, we show that the resulting portfolio is delta-neutral and long volatility.

We construct a weighted portfolio which is *short* the LETF with leverage ratio $\beta_+ > 0$ and *short* another LETF with leverage ratio $\beta_- < 0$. Let us emphasize that both LETFs have the same reference index, but that β_+ and $|\beta_-|$ may differ. A fraction $\omega \in (0, 1)$ of the portfolio is short in the β_+-LETF and $(1-\omega)$ of the portfolio in the β_--LETF. At time T, the return from this strategy is given by

$$\mathcal{R}_T = 1 - \omega \frac{L_T^+}{L_0^+} - (1-\omega)\frac{L_T^-}{L_0^-}. \tag{2.18}$$

Applying (2.17), $\mathcal{R}_T$ can be expressed as an explicit function of the return and realized variance of the reference index. That is,

$$\mathcal{R}_T = 1 - \omega \left(\frac{S_T}{S_0}\right)^{\beta_+} \exp\left(\Gamma_T^+\right) - (1-\omega)\left(\frac{S_T}{S_0}\right)^{\beta_-} \exp\left(\Gamma_T^-\right), \tag{2.19}$$

where

$$\Gamma_T^\pm = \frac{\beta_\pm - \beta_\pm^2}{2} V_T + ((1-\beta_\pm)r - f_\pm)T, \tag{2.20}$$

Here, $\beta_\pm$ and $f_\pm$ are the respective leverage ratios and fees of the two LETFs in the portfolio defined in (2.18). Note that the return $\mathcal{R}_T$ over a short holding period such that $\frac{L_T}{L_0} \approx 1$, one can pick an appropriate weight ω^* to approximately remove the dependence of $\mathcal{R}_T$ on S_T.

If we select the portfolio weight

$$\omega^* = \frac{-\beta_-}{\beta_+ - \beta_-}, \tag{2.21}$$

then the return from this strategy is given by

$$\boxed{\mathcal{R}_T \approx \frac{-\beta_-\beta_+}{2} V_T - \frac{\beta_-}{\beta_+ - \beta_-}(f_+ - f_-)T + (f_- - r)T.} \tag{2.22}$$

To see this, we substitute for $\frac{L_T}{L_0}$ with $\log \frac{L_T}{L_0} + 1$, which is valid when $\frac{L_T}{L_0} \approx 1$. Then, setting $\omega = \frac{-\beta_-}{\beta_+ - \beta_-}$ and applying (2.17), we arrive at (2.22).

The static portfolio return corresponding to the weight ω^* in (2.21) reflects a linear dependence on the realized variance. In particular, the coefficient $\frac{-\beta_-\beta_+}{2}$ in (2.22) is strictly positive, so the strategy is effectively long realized variance. Thus, we call this strategy long volatility. Also, as it does not depend on S_T, the ω^* portfolio is Δ-neutral as long as the reference does not move significantly.

In Table 2.11, we summarize the coefficient of V_T and the (short) portfolio weights $(\omega^*, 1 - \omega^*)$ for different combinations of leverage ratios. Note that as long as $\beta_+ = -\beta_-$, we end up with the portfolio weight $\omega^* = 0.5$, which means we short both $\pm\beta$ LETFs of the same cash amount. All ω^*'s in the table are between 0 and 1, so a short position is taken in both the long and short LETFs. Also, the coefficient $\frac{-\beta_-\beta_+}{2}$ exceeds or equals to 1 except for the pair $(\beta_+, \beta_-) = (1, -1)$. The pair $(\beta_+, \beta_-) = (3, -3)$ corresponds to the most long-volatility portfolio with a coefficient of 4.5.

(β_+, β_-)	ω^*	$\frac{-\beta_-\beta_+}{2}$
$(1, -1)$	$1/2$	$1/2$
$(1, -2)$	$2/3$	1
$(1, -3)$	$3/4$	$3/2$
$(2, -1)$	$1/3$	1
$(2, -2)$	$1/2$	2
$(2, -3)$	$3/5$	3
$(3, -1)$	$1/4$	$3/2$
$(3, -2)$	$2/5$	3
$(3, -3)$	$1/2$	$9/2$

Table 2.11: Table of leverage ratios pairing (β_+, β_-), the static Δ-neutral portfolio weight ω^* (short position for the β_+-LETF), and realized variance coefficient $\frac{-\beta_-\beta_+}{2}$. Note that the (short) weight for the β_--LETF is $(1 - \omega^*)$.

We now backtest the Δ-neutral strategy. For each LETF pair, we short \$0.5 of the β_+-LETF and \$0.5 of the β_--LETF with $\beta_+ = -\beta_- = 2$ and hold the position for 10 days. The return $\mathcal{R}_T$ depends on the relative weights on the long/short-LETFs but not the absolute cash amounts. Dividing the price data from the trading days during 2013–2015 into 10-day rolling periods, we calculate the returns from the strategy over each period. For every 10-day return, we compare it against the realized variance over the same time window.

This is illustrated in Figure 2.7 for the S&P-based LETFs, namely, SSO and SDS with ±2 leverage ratios, and UPRO and SPXU with ±3 leverage ratios. As a theoretical benchmark, we also plot $\mathcal{R}_T$ in (2.22) as a linear function. The 10-day returns are recorded over rolling periods, so they are not independent. In view of this, we emphasize that the straight lines in Figure 2.7 are not generated by regression but taken directly from (2.22). We choose (2.22) as a benchmark because it is expected to hold *pathwise* as long as $\frac{L_T}{L_0} \approx 1$ with negligible tracking error.

We observe from Figure 2.7 that the returns exhibit the predicted positive dependence on the realized variance (V_T) of the reference index. This confirms our intuition that the proposed strategy captures the volatility decay of LETF

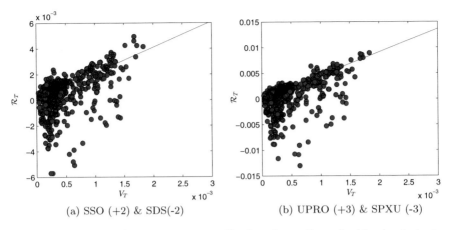

Fig. 2.7: Plot of trading returns vs realized variance for a double short strategy over 5-day rolling holding periods, with $\beta_\pm = \pm 2$ for each LETF pair. We compare with the empirical returns (circle) from the ω^* strategy with the predicted return (solid line) in Prop. (2.22).

as profit. Comparing between the ± 2 LETF pair (SSO & SDS) and ± 3 pair (UPRO & SPXU), the latter is more sensitive to the realized variance. This is because the ± 3 pair has a larger coefficient of V_T in (2.22) than the ± 2 pair (4.5 vs 2 as seen in Table 2.11).

Nevertheless, there is also a visible amount of noise in the returns, especially for the ± 3 pair. The deviation from the straight line is asymmetric in both cases and is particularly skewed to the negative side for the UPRO-SPXU pair. This can be explained by the larger tracking errors commonly experienced by highly leveraged ETFs, such as the ± 3 pair.

While the ω^* portfolio is expected to be Δ-neutral for small movements in the reference index, the strategy is also short-Γ (with respect to the reference index). One way to see this is through Figure 2.8 that plots the returns against the reference index returns. Common to both LETF pairs, when the reference return is either very positive or negative, the return of the ω^*-strategy tends to be negative as a result of the short-Γ property. As a theoretical benchmark, we also plot the normalized return equation (2.19) which does not involve any approximation and applies even for large reference movements under the general diffusion model for the reference index. We see that the empirical returns follow the theoretical benchmark closely and show no directional trend when the reference index return is around zero as a consequence of the portfolio's Δ-neutrality.

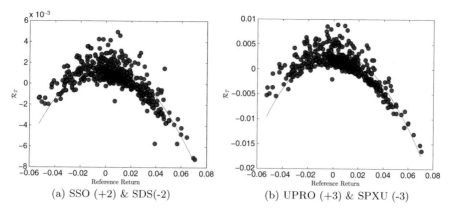

Fig. 2.8: Plot of trading returns vs realized variance for a double short strategy over 5-day rolling holding periods, with $\beta_\pm = \pm 2$ for each LETF pair. We compare with the empirical returns (circle) from the ω^* strategy with the predicted return (solid line) in (2.22).

More generally, one can also test strategies with other leverage ratios $\beta_\pm$ and the corresponding ω^*, or for other non-equity LETFs. For example, we refer to Guo and Leung (2015) for the implementation of this strategy for commodity LETFs.

Chapter 3
Risk Analysis of Leveraged ETFs

In this chapter, we investigate the impact of leverage ratio on the risk associated with an LETF. For simplicity, we assume μ and σ to be constant. From the common risk measures, such as value-at-risk (VaR) and conditional value-at-risk (CVaR), we derive the range of leverage ratios that are admissible with respect to a given risk budget. For each LETF, we analyze the maximum holding period such that the risk constraint is satisfied. As an extension, an intra-horizon risk measure is also studied.

Let us first look at the mean and standard deviation of the discounted relative return of an LETF:

$$\hat{r}(\beta) \equiv \mathbb{E}\left\{L_t/L_0 - 1\right\} = e^{(\beta(\mu-r)+r-f)T} - 1, \tag{3.1}$$

$$\hat{\sigma}(\beta) \equiv \text{std}\left\{L_t/L_0 - 1\right\} = e^{(\beta(\mu-r)+r-f)T}\sqrt{e^{\beta^2\sigma^2 T} - 1}.$$

When selecting the leverage ratio β, a risk-sensitive investor may consider the ratio $\hat{r}(\beta)/\hat{\sigma}(\beta)$ which, loosely speaking, represents the unit of return that one gets for each unit of risk, or the mean-variance trade-off. To choose a range of leverage ratios, the investor can require that

$$\frac{\hat{r}(\beta)}{\hat{\sigma}(\beta)} \geq c, \tag{3.2}$$

where $c > 0$ is the mean-variance trade-off coefficient.

Figure 3.1 shows the mean-variance frontier, along which the leverage ratio varies from 0 to 3. With a positive drift $\mu = 10\%$, negative leverage ratios yield inferior expected return for the same standard deviation, and thus are

© The Author(s) 2016

T. Leung, M. Santoli, *Leveraged Exchange-Traded Funds*, SpringerBriefs
in Quantitative Finance, DOI 10.1007/978-3-319-29094-2_3

not shown in this figure. The mark 'o' (resp. '*') locates the critical leverage ratio β^* satisfying (3.2) in equality with $c = 0.23$ (resp. $c = 0.25$). On the right-hand side of the marks, the ratio $\hat{r}(\beta)/\hat{\sigma}(\beta)$ falls below the required level c, effectively preventing the investor from selecting the higher leverage ratios.

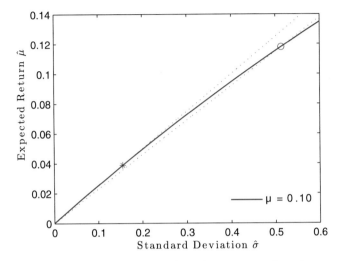

Fig. 3.1: The mean-variance frontier (solid), where each point corresponds to a different leverage ratio within $[0, 3]$. The mark 'o' (resp. '*') locates the critical leverage ratio β^* satisfying (3.2) in equality with $c = 0.23$ (resp. $c = 0.25$).

3.1 Admissible Leverage Ratio

Given a risk budget based on a risk measure, one can determine the leverage ratios that are deemed acceptable. To this end, we now compute analytically the value-at-risk (VaR) and conditional VaR associated with a long position in a leveraged ETF, and derive the associated admissible leverage ratios.

Given a fixed investment horizon T, the probability that the LETF will suffer a relative loss greater than $z \in [0, 1]$ is given by

$$p(z, \beta, T) = \mathbb{P}\{1 - L_T/L_0 > z\} \tag{3.3}$$

$$= \Phi\left(\frac{\log(1 - z) - \psi(\beta)T}{|\beta|\sigma\sqrt{T}}\right), \tag{3.4}$$

where $\Phi(\cdot)$ is the normal cumulative distribution function and

$$\psi(\beta) = \beta(\mu - r) + r - f - \frac{\beta^2 \sigma^2}{2}. \tag{3.5}$$

Note that $\psi(\beta)t$ is related to the log return difference between L and S in (2.7). Figure 3.2 illustrates that the loss probability increases drastically with higher leverage ratios.

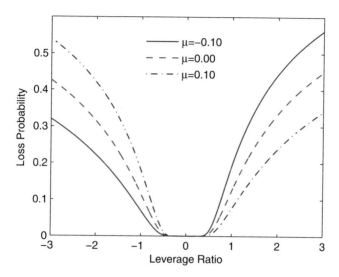

Fig. 3.2: Loss probability increases drastically with higher leverage ratios. Parameters: $z = 20\%$, $r = 2\%$, $T = 0.5$, $\sigma = 25\%$, and $f = 0.95\%$.

In view of the continuous distribution of L_T, we define the (relative) **value-at-risk**, VaR_α, at confidence level $\alpha \in (0,1)$, via the equation

$$p(VaR_\alpha, \beta, T) = \alpha. \tag{3.6}$$

Intuitively, a larger VaR_α means a higher level of risk. For our analysis, $VaR_\alpha \equiv VaR_\alpha(\beta, T)$ is often viewed as a function of leverage ratio β and horizon T. Inverting the loss probability function in (3.6), we obtain an expression for VaR_α. Indeed, given any investment horizon T and leverage ratio β, the (relative) value-at-risk of holding the LETF is given by

$$VaR_\alpha(\beta,T) = 1 - \exp\left(\psi(\beta)T + |\beta|\sigma\sqrt{T}\,\Phi^{-1}(\alpha)\right). \qquad (3.7)$$

with $\psi(\beta)$ defined in (3.5).

To better understand the property of $VaR_\alpha(\beta,T)$, we differentiate w.r.t. β to get

$$\frac{\partial VaR_\alpha}{\partial \beta} = ((\mu - r - \beta\sigma^2)T + \text{sign}(\beta)\Phi^{-1}(\alpha)\sigma\sqrt{T})(VaR_\alpha - 1).$$

Note that this derivative is discontinuous at $\beta = 0$ and changes sign once. In practical applications, the term $\Phi^{-1}(\alpha)\sigma\sqrt{T}$ is negative, and therefore the jump in $\beta = 0$ is upward. Given these observations, either the derivative vanishes at some β^*, or the derivative is negative for $\beta < 0$ and positive for $\beta > 0$. To summarize, we define

$$\beta^* = \begin{cases} \frac{\mu-r}{\sigma^2} + \frac{\Phi^{-1}(\alpha)}{\sigma\sqrt{T}} & \text{if } \frac{\mu-r}{\sigma^2} + \frac{\Phi^{-1}(\alpha)}{\sigma\sqrt{T}} > 0, \\ \frac{\mu-r}{\sigma^2} - \frac{\Phi^{-1}(\alpha)}{\sigma\sqrt{T}} & \text{if } \frac{\mu-r}{\sigma^2} - \frac{\Phi^{-1}(\alpha)}{\sigma\sqrt{T}} < 0, \\ 0 & \text{otherwise,} \end{cases} \qquad (3.8)$$

and conclude that $VaR_\alpha(\beta)$ is decreasing in β for $\beta \leq \beta^*$ and increasing for $\beta \geq \beta^*$. This is illustrated in Figure 3.3. Also, note that β^* does not depend on the expense rate f and increases linearly with the excess return.

One way to describe an investor's risk tolerance is to consider the maximum VaR_α threshold $\bar{z}$ which leads to the inequality in β

$$VaR_\alpha(\beta,T) \leq \bar{z}. \qquad (3.9)$$

This will in turn exclude a range of leverage ratios β. Notice that at $\beta = 0$,

$$VaR_\alpha(0,T) = 1 - e^{(r-f)T},$$

which is typically a low (positive) point of VaR_α, as discussed above and shown in Figure 3.3(a). In Figure 3.3(a), VaR_α is always increasing in $|\beta|$ and $\beta^* = 0$. Moreover, VaR_α is not symmetric in β: VaR_α tends to be lower for those leverage ratios β with the same sign as drift μ. In Figure 3.3(b), we illustrate that, with a higher α, VaR_α reaches its minimum at $\beta^* \neq 0$.

Therefore, if risk tolerance is very low, or if the expense ratio is high, there might not exist any leverage ratio β for which the investor is willing to invest. The admissible range of leverage ratios based on criterion (3.9) is given by

$$I^- \cup I^+,$$

where

$$I^\pm = \begin{cases} [l^\pm, u^\pm] & \text{if } \Delta^\pm \text{ is real,} \\ \emptyset & \text{otherwise,} \end{cases}$$

with

$$l_+ = \max\{0, \Gamma^+(\alpha) - \Delta^+(\alpha)\}, \quad u_+ = \max\{0, \Gamma^+(\alpha) + \Delta^+(\alpha)\}, \quad (3.10)$$
$$l_- = \min\{0, \Gamma^-(\alpha) - \Delta^-(\alpha)\}, \quad u_- = \min\{0, \Gamma^-(\alpha) + \Delta^-(\alpha)\}, \quad (3.11)$$

and

$$\Gamma^\pm(\alpha) = \frac{1}{\sigma^2 T}\left((\mu - r)T \pm \Phi^{-1}(\alpha)\sigma\sqrt{T}\right),$$
$$\Delta^\pm(\alpha) = \frac{1}{\sigma^2 T}\sqrt{((\mu - r)T \pm \Phi^{-1}(\alpha)\sigma\sqrt{T})^2 - 2\sigma^2 T((f - r)T + \log(1 - \bar{z}))}.$$

As a result, we can identify precisely the interval of acceptable leverage ratios. To visualize this, we look at Figure 3.3(a) and consider the leverage ratios whose value-at-risk is lower than a given level. For instance, setting the risk tolerance level $\bar{z} = 0.5$, the leverage ratios 2 and -2 are admissible but 3 and -3 are excluded. In Figure 3.3, setting $\bar{z} = 0.25$, the admissible leverage ratio interval for $\mu = -18\%$ is $[l^-, u^-] \cup [l^+, u^+] = [-2.99, 1.21]$, so $\beta = -3$ is excluded. On the other hand, with $\mu = 18\%$, the admissible interval is $[-1.26, 2.76]$, so $\beta = 3$ is excluded. The concept of admissible leverage ratio provides a simple recipe for identifying LETFs that are too risky according to a given risk measure.

In addition to value-at-risk, we define the **conditional value-at-risk** $CVaR_\alpha$ at confidence level α as

$$CVaR_\alpha(\beta, T) := \mathbb{E}\left\{1 - \frac{L_T}{L_0} \,\middle|\, 1 - \frac{L_T}{L_0} > VaR_\alpha(\beta, T)\right\}.$$

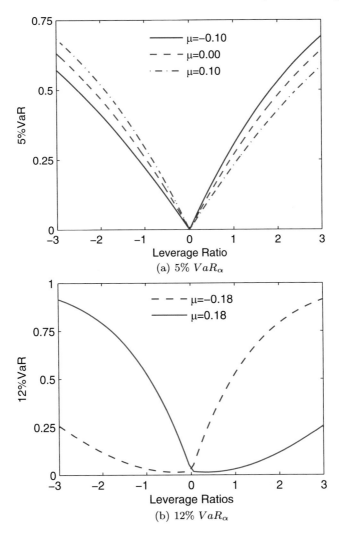

(a) 5% VaR_α

(b) 12% VaR_α

Fig. 3.3: (a) VaR_α is lowest at $\beta = 0$ and increases as the absolute value of leverage ratio $|\beta|$ increases. Parameters: $r = 2\%, T = 0.5$, $\sigma = 25\%$, and $f = 0.95\%$. (b) VaR_α achieves a minimum at a nonzero leverage, i.e., $\beta^* \neq 0$ (see (3.8)). Parameters: $r = 0\%, T = 2$, $\sigma = 20\%$, and $f = 0.95\%$.

Given any investment horizon T and leverage ratio β, the conditional value-at-risk for the LETF is given by

$$CVaR_\alpha(\beta, T) = 1 - e^{(\beta(\mu-r)+r-f)T} \frac{\Phi(\Phi^{-1}(\alpha) - |\beta|\sigma\sqrt{T})}{\alpha}. \qquad (3.12)$$

The $CVaR_\alpha(\beta)$ is decreasing in β for $\beta \leq \beta^{**}$ and increasing for $\beta \geq \beta^{**}$, with the critical leverage β^{**} satisfying

$$\frac{\mu - r}{\sigma\sqrt{T}}\Phi(\Phi^{-1}(\alpha) - |\beta^{**}|\sigma\sqrt{T}) = sign(\beta^{**})\phi(\Phi^{-1}(\alpha) - |\beta^{**}|\sigma\sqrt{T}). \quad (3.13)$$

To see this, for any $\beta \in \mathbb{R}$, we compute the expected (relative) shortfall explicitly as

$$\mathbb{E}\left\{\frac{L_0 - L_T}{L_0} \mid \frac{L_0 - L_T}{L_0} > z\right\} = 1 - e^{(\beta(\mu-r)+r-f)T}\frac{\Phi(d_z - |\beta|\sigma\sqrt{T})}{\Phi(d_z)}, \qquad (3.14)$$

where

$$d_z := \frac{\log(1 - z) - \psi(\beta)T}{|\beta|\sigma\sqrt{T}}.$$

Putting $z = VaR_\alpha$ gives $CVaR_\alpha$ in (3.12). Next, we compute the derivative

$$\frac{\partial CVaR_\alpha}{\partial \beta} = \frac{e^{(\beta(\mu-r)+r-f)T}}{\alpha} \times$$

$$\left((r - \mu)\Phi(\Phi^{-1}(\alpha) - |\beta|\sigma\sqrt{T}) + sign(\beta)\sigma\sqrt{T}\phi(\Phi^{-1}(\alpha) - |\beta|\sigma\sqrt{T})\right).$$

The sign of the derivative depends on the term in the bracket, and equating this to zero yields the critical value β^{**} in (3.13).

3.2 Admissible Risk Horizon

The risk analysis in the previous section sheds light on the choice of leverage ratios. Alternatively, the investor can control risk exposure by appropriately selecting the investment horizon. For risk management purposes, it is important to determine the maximum investment horizon τ such that the risk measure stays under some threshold $C \in (0, 1)$. This idea leads us to study the *admissible risk horizon* induced by a risk measure.

First, let us consider the value-at-risk $VaR_\alpha(\beta,\tau)$ for a β-LETF and horizon τ. The **admissible risk horizon** $ARH_\alpha(\beta,C)$ is defined by

$$ARH_\alpha(\beta,C) = \inf\left\{\tau \geq 0 : VaR_\alpha\left(\beta,\tau\right) = C\right\}, \qquad (3.15)$$

and we set $ARH_\alpha(\beta,C) = +\infty$ if the equation $VaR_\alpha\left(\beta,\tau\right) = C$ has no positive root (in τ). Herein, we impose an upper bound of 0.5 on α so that $\Phi(\alpha) < 0$. Using formula (3.7), we invert (3.15) to get an explicit expression for $ARH_\alpha(\beta,C)$.

Denote $b = |\beta|\,\sigma\Phi^{-1}(\alpha)$. If $b^2/4 \geq -\psi(\beta)\log\left(1-C\right)$, then the admissible risk horizon for the β-LETF with VaR_α limited at C is given by

$$ARH_\alpha(\beta,C) = \left(\frac{-b/2 - \sqrt{b^2/4 + \psi(\beta)\log\left(1-C\right)}}{\psi(\beta)}\right)^2. \qquad (3.16)$$

If $b^2/4 < -\psi(\beta)\log\left(1-C\right)$, then $ARH_\alpha(\beta,C) = +\infty$.

To gain some insight on the conditions for (3.16), we first recall that

$$VaR_\alpha(\beta,T) = 1 - \exp\left(\psi(\beta)T + |\beta|\sigma\sqrt{T}\,\Phi^{-1}(\alpha)\right).$$

If, given a certain β, $\psi(\beta) > 0$, then VaR_α is convex and eventually decreasing in T, and its maximum can potentially lie below the threshold C. In fact, whether this happens or not is determined by the condition $b^2/4 < -\psi(\beta)\log\left(1-C\right)$. In reality, we tend to have $\psi(\beta) > 0$ when β is small or when $\text{sign}(\beta)\mu$ is large. The latter means that the investment has a high rate of return, so intuitively VaR_α can stay low.

On the other hand, when $b^2/4 > -\psi(\beta)\log\left(1-C\right)$ and $\psi(\beta) > 0$, the equation $VaR_\alpha\left(\beta,\tau\right) = C$ admits two positive roots and (3.16) selects only the smallest root, according to equation (3.15).

In contrast, if $\psi(\beta) < 0$, then VaR_α is increasing in T. Consequently, the equation $VaR_\alpha\left(\beta,\tau\right) = C$ always admits a unique strictly positive solution (note that $VaR_\alpha\left(\beta,0\right) = 0$).

Figure 3.4 illustrates how ARH_α varies for different values of β and μ. As we can see, the admissible risk horizon increases as $|\beta|$ decreases. For any fixed positive leverage ratio, the admissible risk horizon tends to increase with drift μ. In addition, it is also preferable to choose a leverage ratio, say $\hat\beta$, with the same sign as that of μ, since the corresponding admissible risk horizon is greater than that of $-\hat\beta$.

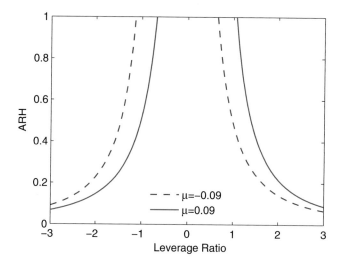

Fig. 3.4: The admissible risk horizon tends to decrease as leverage ratio deviates from zero. Parameters: $C = 0.25, r = 1\%, \alpha = 5\%, \sigma=25\%$, and $f = 0.95\%$.

Similarly, one can define an admissible risk horizon based on other risk measures. For instance, the admissible risk horizon $\widehat{ARH}_\alpha(\beta, C)$ based on the conditional value-at-risk is determined from the equation:

$$\widehat{ARH}_\alpha(\beta, C) = \inf \left\{ \tau \geq 0 \, : \, CVaR_\alpha\left(\beta, \tau\right) = C \right\}. \qquad (3.17)$$

In this case an analytical solution is not available. Nevertheless, one can easily find the zero(s) of the function $g(\tau) := CVaR_\alpha(\beta, \tau) - C$.

3.3 Intra-Horizon Risk and Stop-Loss Exit

Both VaR and CVaR concern the loss distribution at a fixed future date, even though the LETF value may experience large losses at intermediate times. In reality, investors may monitor the asset price movement and impose a stop-loss level to limit downside risk. This motivates us to model a stochastic holding period until the LETF falls to a certain threshold.

We define the first passage time that the LETF, starting at L_0, reaches to a lower level ℓL_0:

$$\tau_\ell = \inf\{\, t \geq 0 \,:\, L_t \leq \ell L_0 \,\}, \qquad \ell \in (0,1).$$

With this, we define the intra-horizon loss probability

$$\underline{p}(\ell, \beta, T) = \mathbb{P}\{\tau_\ell \leq T\}.$$

This probability is related to the minimum of L over $[0,T]$, denoted by $\underline{L}_T = \min_{0 \leq t \leq T} L_t$, and admits an explicit expression:

$$\underline{p}(\ell, \beta, T) = \mathbb{P}\{\underline{L}_T \leq \ell L_0\}$$
$$= \Phi\left(\frac{\log(\ell) - \psi(\beta)T}{|\beta|\,\sigma\sqrt{T}}\right) + \ell^{2\psi(\beta)/(\beta\sigma)^2} \Phi\left(\frac{\log(\ell) + \psi(\beta)T}{|\beta|\,\sigma\sqrt{T}}\right),$$

$$(3.18)$$

where $\psi(\beta)$ is defined in (3.5).

Figure 3.5 (top) shows how $\underline{p}(\ell, \beta)$ varies with respect to β for three different threshold levels. As ℓ or β increases, the intra-horizon loss probability $\underline{p}(\ell, \beta)$ increases. In addition, for any pair of leverage ratios with the same absolute value, $\underline{p}(\ell, \beta)$ is lower for the positive leverage ratio.

Given any loss level ℓL_0, it is useful to know whether an LETF will reach it in the long run. This amounts to determining whether the probability $\mathbb{P}\{\tau_\ell < \infty\}$ is equal to or strictly less than 1. Taking $T \uparrow \infty$ in (3.18), direct computation shows that, when $\psi(\beta) \leq 0$, we have $\mathbb{P}\{\tau_\ell < \infty\} = 1$. In other words, to ensure $\mathbb{P}\{\tau_\ell < T\} < 1$, one has to restrict the leverage choice so that $\psi(\beta) > 0$.

To understand the condition in terms of leverage ratio β, we recall from (3.5) that $\psi(\beta)$ is not always of constant sign, and is quadratic in leverage ratio β. In particular, for large $|\beta|$, the volatility σ will have a dominant negative impact. If $f \leq r$, then we have $\psi(\beta) > 0$ over the interval of leverage ratios:

$$\beta \in \left(\frac{\mu - r - \sqrt{(\mu - r)^2 + 2\sigma^2(r - f)}}{\sigma^2}, \frac{\mu - r + \sqrt{(\mu - r)^2 + 2\sigma^2(r - f)}}{\sigma^2}\right).$$

Note that the width of the leverage interval is proportional to the excess return $\mu - r$ and inversely proportional to volatility. On the other hand, if

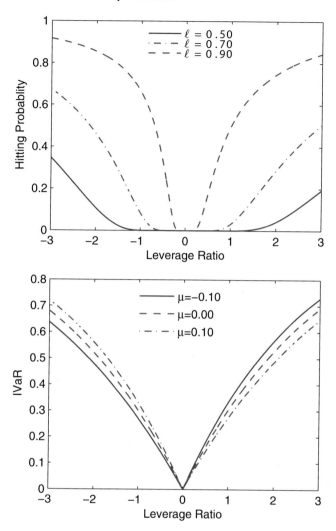

Fig. 3.5: Top: Probability of hitting a lower level ℓ for different leverage ratios in [-3, 3], with $\mu = 10\%$. Bottom: Intra-horizon VaR vs leverage ratios, with $\alpha = 5\%$. Common parameters for both plots: $r = 2\%$, $T = 0.5$, $\sigma = 25\%$, and $f = 0.95\%$.

$f > r$, then the above interval holds if and only if $(\mu - r)^2 \geq 2\sigma^2(f - r)$. Otherwise, we have $\psi(\beta) \leq 0$ and the interval of leverage ratio does not exist. In this case, the hitting time τ_ℓ is finite almost surely.

The intra-horizon loss probability leads us to define the *intra-horizon value-at-risk*, denoted by $IVaR_\alpha$, via the equation

$$\underline{p}\big(1 - IVaR_\alpha, \beta, T\big) = \alpha. \tag{3.19}$$

This is a modification of value-at-risk, and it incorporates the possibility that the LETF *ever* falls below a lower level. Due to the complicated form of equation (3.18), this risk metric does not admit an explicit formula. Nevertheless, the numerical solution for $IVaR_\alpha$ in (3.19) involves a straightforward and instant root finding.

In Figure 3.5 (bottom), we illustrate how $IVaR_\alpha$ varies with respect to β for different values of μ. Similar to VaR_α, $IVaR_\alpha$ also increases as leverage ratio deviates from 0. Notice that with common parameters, we have $IVaR_\alpha (\beta, T) > VaR_\alpha (\beta, T)$. This can be inferred from the definitions (3.6) and (3.19), along with the fact that $\underline{p}(1 - z, \beta, T) > p(z, \beta, T)$ (see (3.4) and (3.18)).

The intra-horizon value-at-risk admits the partial derivative

$$\frac{\partial IVaR_\alpha}{\partial \beta} = \frac{(1 - IVaR_\alpha)\,|\beta|\,\sigma\,\frac{\partial \underline{p}}{\partial \beta}}{\frac{1}{\sqrt{T}}\phi(\hat{d}_-) + \frac{2\psi(\beta)}{|\beta|\sigma}(1 - IVaR_\alpha)^{2\psi(\beta)/(\beta\sigma)^2}\Phi(\hat{d}_+) + \frac{(1-IVaR_\alpha)^{2\psi(\beta)/(\beta\sigma)^2}}{\sqrt{T}}\phi(\hat{d}_+)}, \tag{3.20}$$

where

$$\hat{d}_\pm := \frac{\log(1 - IVaR_\alpha) \pm \psi(\beta)T}{|\beta|\,\sigma\sqrt{T}}.$$

We can deduce from (3.20) that the derivative would take the same sign as $\frac{\partial \underline{p}(z,\beta)}{\partial \beta}$, unless the drift term $\psi(\beta)$ is so negative that the denominator also becomes negative.

The intra-horizon risk measure motivates a stop-loss exit strategy in order to limit downside risk during the investment horizon. Incorporating a stop-loss level ℓL_0, we denote $\mathcal{R}_T = \frac{L_{T\wedge\tau_\ell}}{L_0}$ and express the expected relative value as

$$\mathbb{E}\{\mathcal{R}_T\} = \ell\,\mathbb{P}\{\tau_\ell < T\} + \mathbb{E}\left\{\frac{L_T}{L_0}1_{\{\tau_\ell > T\}}\right\}. \tag{3.21}$$

The first term is given by

$$\ell \mathbb{P}\{\tau_\ell < T\} = \ell \left[\Phi \left(\frac{\log(\ell) - \psi(\beta)T}{|\beta|\sigma\sqrt{T}} \right) + \ell^{2\psi(\beta)/(\beta\sigma)^2} \Phi \left(\frac{\log(\ell) + \psi(\beta)T}{|\beta|\sigma\sqrt{T}} \right) \right].$$

For the second term, we apply standard calculations to get

$$\mathbb{E}\left\{ \frac{L_T}{L_0} 1_{\{\tau_\ell > T\}} \right\} = e^{(\psi(\beta) + \frac{\beta^2\sigma^2}{2})T} \left[\Phi \left(|\beta|\sigma\sqrt{T} + \frac{\psi(\beta)\sqrt{T}}{|\beta|\sigma} - \frac{\log(\ell)}{|\beta|\sigma\sqrt{T}} \right) \right.$$
$$\left. - \ell^{2\psi(\beta)/(\beta\sigma)^2 + 2} \Phi \left(|\beta|\sigma\sqrt{T} + \frac{\psi(\beta)\sqrt{T}}{|\beta|\sigma} + \frac{\log(\ell)}{|\beta|\sigma\sqrt{T}} \right) \right].$$
$$\tag{3.22}$$

Moreover, the variance of the expected relative value with a stop-loss exit also admits a closed-form formula. Precisely, the variance is given by $\mathrm{var}\{\mathcal{R}_T\} = \mathbb{E}\{\mathcal{R}_T^2\} - (\mathbb{E}\{\mathcal{R}_T\})^2$, with $\mathbb{E}\{\mathcal{R}_T\}$ from (3.21) and

$$\mathbb{E}\{\mathcal{R}_T^2\} = \frac{1}{L_0^2} \mathbb{E}\left\{ \ell^2 1_{\{\tau_\ell < T\}} + L_T^2 1_{\{\tau_\ell > T\}} \right\}$$
$$= \ell^2 \left[\Phi \left(\frac{\log(\ell) - \psi(\beta)T}{|\beta|\sigma\sqrt{T}} \right) + \ell^{2\psi(\beta)/(\beta\sigma)^2} \Phi \left(\frac{\log(\ell) + \psi(\beta)T}{|\beta|\sigma\sqrt{T}} \right) \right]$$
$$+ e^{2T(\psi(\beta) + \beta^2\sigma^2)} \left[\Phi \left(2|\beta|\sigma\sqrt{T} + \frac{\psi(\beta)\sqrt{T}}{|\beta|\sigma} - \frac{\log(\ell)}{|\beta|\sigma\sqrt{T}} \right) \right.$$
$$\left. - \ell^{2\psi(\beta)/(\beta\sigma)^2 + 4} \Phi \left(2|\beta|\sigma\sqrt{T} + \frac{\psi(\beta)\sqrt{T}}{|\beta|\sigma} + \frac{\log(\ell)}{|\beta|\sigma\sqrt{T}} \right) \right].$$

Suppose the investor also seeks to take profit when the asset reaches a sufficiently high level. Then it is useful to compute the probability that LETF will fall to the stop-loss level before reaching the take-profit level. To fix ideas, we denote τ_ℓ (resp. τ_h) as the time for the LETF to hit a lower level ℓL_0 (resp. upper level hL_0), with $\ell \leq 1 \leq h$.

The probability that a β-LETF to reach a lower level ℓL_0 before a higher level hL_0 is given by

$$\mathbb{P}\{\tau_\ell < \tau_h\} = \frac{1 - h^{-2\psi(\beta)/(\beta\sigma)^2}}{\ell^{-2\psi(\beta)/(\beta\sigma)^2} - h^{-2\psi(\beta)/(\beta\sigma)^2}}.$$
$$\tag{3.23}$$

The explicit loss probability formula (3.23) can be used to quantify the risk of holding an LETF with a take-profit/stop-loss strategy. A risk-sensitive investor may want to bound it by $q \in (0,1)$, namely,

$$\mathbb{P}\{\tau_\ell < \tau_h\} \le q. \tag{3.24}$$

Let us think of the loss probability bound in (3.24) as exogenously specified. The investor selects the maximum admissible take-profit level h_{max} in terms of stop-loss level ℓ and probability bound q. From (3.23), we observe that $\mathbb{P}\{\tau_\ell < \tau_h\}$ monotonically increases with h, and arrive at two cases:

- Case 1: $\psi(\beta) > 0$ and $q \ge \ell^{2\psi(\beta)/(\beta\sigma)^2}$.
  As $h \to \infty$, $\mathbb{P}\{\tau_\ell < \tau_h\}$ is bounded by $\ell^{2\psi(\beta)/(\beta\sigma)^2} \le 1$. If $q \ge \ell^{2\psi(\beta)/(\beta\sigma)^2}$, then the take-profit level can be arbitrarily high by the investor, i.e., $h_{max} = \infty$.

- Case 2: $\psi(\beta) > 0$ and $q < \ell^{2\psi(\beta)/(\beta\sigma)^2}$, or $\psi(\beta) < 0$.
  From (3.23), $h = 1 \Rightarrow \mathbb{P}\{\tau_\ell < \tau_h\} = 0$. By monotonicity, there exists a finite upper level $h_{max} \in [1,\infty)$. By rearranging the inequality (3.24), we obtain the maximum admissible take-profit level

$$h_{max} = \left(\frac{1 - \ell^{-2\psi(\beta)/(\beta\sigma)^2}q}{1-q}\right)^{-\frac{\beta^2\sigma^2}{2\psi(\beta)}}. \tag{3.25}$$

From this result, we also infer that h_{max} decreases with ℓ and increases with q.

Chapter 4
Options on Leveraged ETFs

The popularity of ETFs has also led to increased trading of options written on ETFs. In 2012, the total options trading volume on the Chicago Board Options Exchange (CBOE) was 1.06 billion contracts, of which 282 million were ETF options while 473 million were stock options. As we have seen, LETFs that are referenced to the same underlying index share the same source of randomness. This leads to an important question of consistent pricing of options written on LETFs with the same reference index. In other words, we will study the price relationships among LETF options, not only across strikes and maturities but also for various leverage ratios.

In this chapter, we look at the returns of LETF options based on the S&P 500 index. Using empirical prices and realized payoffs of these LETF options, we compare the return distributions across different leverage ratios, ranging from -3 to $+3$. Another way to examine the price relationship among LETF options is through the well-known Put-Call Parity. To this end, we compare the prices of market-traded LETF calls (respectively puts) with the values of synthetic calls (respectively synthetic puts). The observed price discrepancy can be measured by the concept of *implied dividend* and compared across strikes, maturities, as well as leverage ratios.

4.1 Empirical Returns of LETF Options

Traded options have different strikes and expiration dates. For each expiration date, we can compare returns of LETF calls and puts with different strikes and leverage ratios. It is common to consider option prices as a function of

T. Leung, M. Santoli, *Leveraged Exchange-Traded Funds*, SpringerBriefs in Quantitative Finance, DOI 10.1007/978-3-319-29094-2_4

moneyness, which is defined by the ratio of the strike over the spot LETF price. As such, at-the-money options have moneyness of value 1. The prices of options on LETFs with different leverage ratios are generally not comparable at the same strike or moneyness. The reason is that LETFs can differ greatly in their dynamics and dependence on the reference index. For example, an out-of-the-money (OTM) put on a long ETF or LETF ($\beta \geq 1$) is a bearish position (on the reference index) with a low moneyness, while an OTM put on a short LETF ($\beta \leq -1$) also has a low moneyness but is a bullish position on the reference index.

The availability of multiple leverage ratios for the LETF options leads us to investigate the link between these markets. More specifically, compared to a non-leveraged ETF option with moneyness m, what is the comparable moneyness m_β among the LETF option with a given leverage ratio β? In other words, which pair of options is essentially taking the same bet (or market view) on the reference index, measured in terms of expected returns? In order to address these issues and compare LETF options across leverage ratios, we consider the idea of scaling the moneyness of LETF options to make them all comparable with their non-leveraged counterparts.

For each LETF option with a given leverage ratio β and moneyness m' (i.e., K'/L_0 where K' is the option's strike price), we define the *adjusted moneyness* by

$$m_\beta := (m')^{\frac{1}{\beta}} \exp\left(\frac{1}{\beta}(r(\beta - 1) + f)T + \frac{\beta - 1}{2}\bar{\sigma}^2 T\right), \qquad (4.1)$$

where T is the time to maturity of the option, and f is the expense ratio charged by the β-LETF, and $\bar{\sigma}$ is the averaged implied volatility across all available strikes. It is important to note that for short LETFs ($\beta < 0$), the moneyness is inverted approximately around 1, after the adjustment. In other words, for a short LETF OTM and ITM options, we have the adjusted moneyness $m_\beta \gtrless 1$ when the original one $m' \lessgtr 1$. This result echoes the intuition that puts on short LETFs and calls on long LETFs are both bullish, while calls on short LETFs and puts on long LETFs are both bearish. Inverting the moneyness for short LETF bullish/bearish options means that we can align them with the bullish/bearish counterparts written on long LETFs.

For both long and short LETF options, the adjusted moneyness helps convert the moneyness of LETF options into the "same scale" as that of the (non-leveraged) ETF options written on the reference index. To see how this works, we look at the empirical returns of LETF options, which are defined as

$$R_c(m', T) = \frac{(L_T - m'L_0)^+}{C(m'L_0, T)} - 1, \qquad \text{(call)} \qquad (4.2)$$

$$R_p(m', T) = \frac{(m'L_0 - L_T)^+}{P(m'L_0, T)} - 1 \qquad \text{(put)}, \qquad (4.3)$$

where $C(m'L_0, T)$ and $P(m'L_0, T)$ are, respectively, the observed market prices of the LETF call and put with moneyness m' (or equivalently, strike $K' = m'L_0$) and maturity T. Also, L_0 and L_T are, respectively, the spot and terminal prices of the LETF. Next, we plot the average returns of the LETF options over the same x-axis using the adjusted moneyness in (4.1).

In Figure 4.1, we consider all the bullish LETF options (with respect to the reference index) of the S&P 500 LETF series, based on all 1-month options from January 2010 to January 2013. We plot their average returns over their adjusted moneyness. In other words, each data point corresponds to a single adjusted moneyness for the corresponding LETF call/put. As we can see, the returns match very well across leverage ratios and option types. Here, the OTM calls (with high adjusted moneyness) for the long LETFs tend to expire worthless in one month and yield a return of -1, while ITM calls with low adjusted moneyness are more likely to have a positive return; see $+1, +2, +3$ calls in the figure. For short LETFs, the moneyness is inverted, so those puts with an adjusted moneyness greater than 1 correspond to the original OTM LETF puts, and these options tend to expire worthless upon expiration.

Figure 4.2 illustrates the average returns of the 1-month LETF options that are bearish with respect to the S&P 500, recorded from January 2010 to January 2013. Again, plotting the returns over adjusted moneyness shows that the average returns line up well across leverage ratios and option types. For the long LETFs, the puts with high adjusted moneyness tend to expire in the money in one month while OTM puts tend to yield a return of -1. As for the calls on the short LETFs ($-1, -2, -3$ calls in the figure) that have a low adjusted moneyness, they correspond to the original OTM calls and are thus more likely to expire worthless.

We remark that aligning the average option returns by adjusting moneyness does not mean that the option payoffs have the same distribution. Indeed, LETFs with different leverage ratios do not share the same dependence on the reference index. In particular, they have different exposures to volatility decay. Figure 4.3 shows the empirical distributions of the returns of the 1-month bullish ATM options. While they all have a similar profile, the long LETFs, SSO and UPRO, appear to have a more right-skewed distribution of returns. In Figure 4.4, we see similar return distributions among the

1-month bearish ATM options. Here, we see that the SPXU ($\beta = +3$) call has about a 50% chance to expire worthless yielding a return of -1, and this probability is significantly higher than other bearish ATM options.

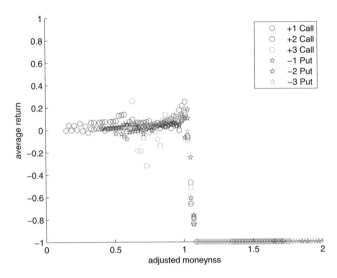

Fig. 4.1: Empirical average returns for the 1-month LETF options (calls on the long LETFs and puts on the short LETFs) that are bullish with respect to S&P 500, plotted against adjusted moneyness.

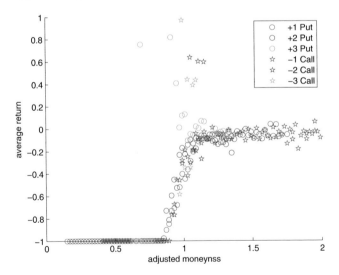

Fig. 4.2: Empirical average returns for the 1-month LETF options (puts on the long LETFs and calls on the short LETFs) that are bearish with respect to S&P 500, plotted against adjusted moneyness.

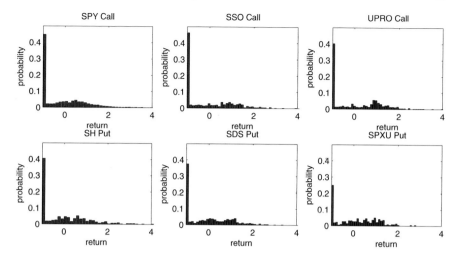

Fig. 4.3: Empirical distributions of the returns for the 1-month ATM LETF options that are bullish with respect to the reference index (S&P500) over the period 1/2010–1/2013. The (L)ETF symbols {SPY, SSO, UPRO, SH, SDS, SPXU} correspond to the leverage ratios $\{+1, +2, +3, -1, -2, -3\}$.

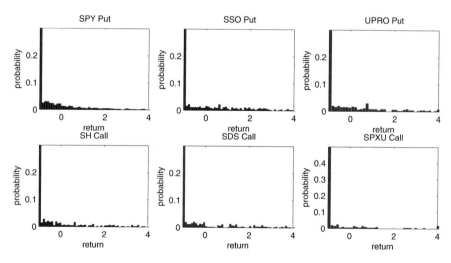

Fig. 4.4: Empirical distributions of the returns for the 1-month ATM LETF options that are bearish with respect to the reference index (S&P500) over the period 1/2010–1/2013. The (L)ETF symbols {SPY, SSO, UPRO, SH, SDS, SPXU} correspond to the leverage ratios $\{+1, +2, +3, -1, -2, -3\}$.

The well-known Put-Call Parity connects the prices of a call and a put with the same underlying, strike and maturity. We can apply this idea to construct synthetic calls and synthetic puts written on LETFs. Since in theory the LETF call (or put) and its synthetic counterpart should have the same price for every strike and maturity, this exercise can reveal the mispricing between LETF puts and calls. It is also practically useful. If a trader wants to take a long position in an LETF call (or put), then it is only natural to buy the cheaper of the two: the LETF call (or put) or its synthetic counterpart.

For a given strike K and maturity T, we denote by $C^b(K,T)$ and $C^a(K,T)$, respectively, the bid and ask prices of the call written on an LETF L. Similarly, $P^b(K,T)$ and $P^a(K,T)$ are the bid and ask put prices, and L^b and L^a are the bid and ask spot LETF prices. By Put-Call Parity, a synthetic call, denoted by $\tilde{C}(K,T)$, can be constructed by borrowing Ke^{-rT} and buying a put and a share of the underlying asset. Similarly, a synthetic put, denoted by $\tilde{P}(K,T)$, involves buying a call, shorting one share of the underlying asset and putting Ke^{-rT} in the money market account. In summary, the long synthetic call and long synthetic put values are given by

$$\tilde{C}(K,T) = P^a(K,T) + L^a - Ke^{-rT},$$
$$\tilde{P}(K,T) = C^a(K,T) - L^b + Ke^{-rT}.$$

We remark that a synthetic call does not involve shorting the LETF or options, but embedded in a synthetic put is a short position in the LETF. The next question is whether the synthetic call and put values coincide with the corresponding long LETF options' prices.

In Figure 4.5, we consider options written on the S&P500 based (L)ETFs {SPY, SSO, UPRO, SH, SDS, SPXU}, respectively, with the leverage ratios $\{+1, +2, +3, -1, -2, -3\}$, and compare the prices of LETF calls and synthetic calls with the same strike and maturity (1 month) by plotting them against moneyness. Notice that the SPY options are the most expensive, and the options written on SH, SDS, or SPXU cost significantly less. This is mainly due to the lower per-unit prices of these LETFs compared to SPY, and does not imply a lower volatility for these LETFs. For all LETFs, the *synthetic* call prices are visibly higher than the original LETF call prices. As one may expect, synthetic options should be priced higher than or equal to the original options. Otherwise, the trading volume of original options would be zero. Therefore, the observations in Figure 4.5 confirms our intuition.

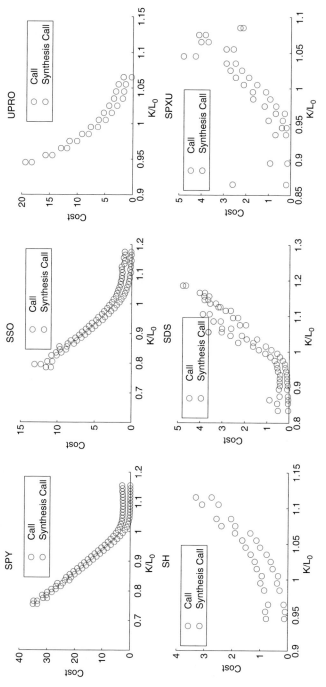

Fig. 4.5: LETF call vs synthetic call prices across moneyness and leverage ratios. All options, taken from January 2012 to January 2013, have 1 month to maturity, and the 1-month LIBOR rate $r = 0.19\%$ is used. The (L)ETF symbols {SPY, SSO, UPRO, SH, SDS, SPXU} correspond to the leverage ratios $\{+1, +2, +3, -1, -2, -3\}$.

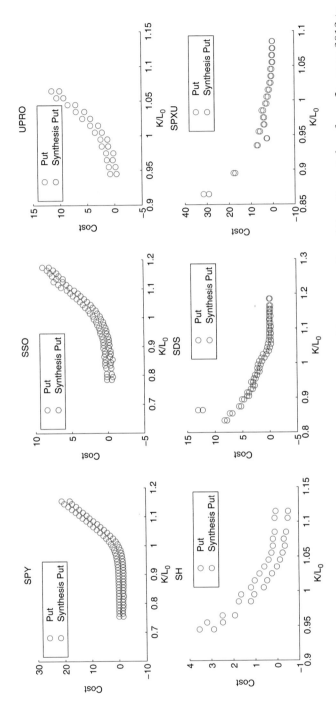

Fig. 4.6: LETF put vs synthetic put prices across moneyness and leverage ratios. All options, taken from January 2012 to January 2013, have 1 month to maturity, and the 1-month LIBOR rate $r = 0.19\%$ is used. The (L)ETF symbols {SPY, SSO, UPRO, SH, SDS, SPXU} correspond to the leverage ratios $\{+1, +2, +3, -1, -2, -3\}$.

However, Figure 4.6 reveals a very different phenomenon. For each strike, the synthetic put price is lower than the corresponding LETF put price, and this is observed for all the 1-month options across all six cases. In contrast to synthetic calls, all synthetic puts require shorting the LETF. We believe this is the primary factor that contributes to lower synthetic put prices. The lower synthetic put prices may suggest that traders should buy them instead of the original LETF puts. In order to do so, the traders need to borrow the underlying LETF. This may involve additional costs and sometimes borrowing the LETF may be infeasible.

4.2 Implied Dividend

In practice, hard-to-borrow (L)ETFs are subject to buy-ins, in the sense that their short positions could be closed at market prices unexpectedly by the clearing firm, along with additional margins and brokerage fees. As a consequence, a long put position and the corresponding synthetic position that involves shorting the underlying are different. Avellaneda and Lipkin (2009) give a detailed account on *hard-to-borrow* stocks and introduce the concept of *implied dividend* to measure the market friction due to short selling, though they do not study the implied dividends associated with LETF options. Here, we adopt this idea for analyzing and comparing LETF options. The implied dividend q is defined through the Put-Call Parity equation:

$$C(K,T) - P(K,T) = Le^{-qT} - Ke^{-rT}.$$

The cost of maintaining a short position is reflected in terms of the implied dividend as if it is paid out by the underlying LETF. Most commonly, LETFs do not pay dividends. Thus, the implied dividend summarizes all costs that come from short-selling. As such, one would expect it to be positive.

In order to compute the implied dividend, we compare the prices of calls and synthetic calls, and also puts and synthetic puts. Consider again the following strategies of zero theoretical portfolio values :

1. Short 1 call, long 1 put, long e^{-qT} share of LETF, borrow Ke^{-rT} in cash;
2. Long 1 call, short 1 put, short e^{-qT} share of LETF, deposit Ke^{-rT} in cash.

Accounting for the bid-ask spread, we have the following cash flows received at time 0:

$$C^b(K,T) - P^a(K,T) - Le^{-qT} + Ke^{-rT}, \qquad (4.4)$$

and

$$- C^a(K,T) + P^b(K,T) + Le^{-qT} - Ke^{-rT}. \tag{4.5}$$

We define the implied dividend by setting (4.4) and (4.5) to zero. Therefore, we obtain

$$q^b := -\frac{1}{T} \log \left(\frac{C^b(K,T) - P^a(K,T) + Ke^{-rT}}{L^a} \right), \tag{4.6}$$

$$q^d := -\frac{1}{T} \log \left(\frac{C^a(K,T) - P^b(K,T) + Ke^{-rT}}{L^b} \right). \tag{4.7}$$

Here, q^b is the implied dividend rate corresponding to the first strategy which involves borrowing cash, and q^d is the implied dividend rate corresponding to the second strategy. Note that for a fixed risk-free rate r, q^b is always higher than q^d. If there is no arbitrage, we must have (4.4) and (4.5) less than or equal to zero. Hence, the no-arbitrage interval for implied dividend rate is $[q^d, q^b]$.

Figure 4.7 gives a snapshot of the implied dividend as a function of moneyness for different maturities. The implied dividends are derived from SPY ($\beta = 1$) options on November 26, 2012. For every maturity, we see that the implied dividend is increasing in moneyness. However, the implied dividend curves become flatter for longer time-to-maturity. As a result, the implied dividend of options with high moneyness tends to decrease as time-to-maturity increases.

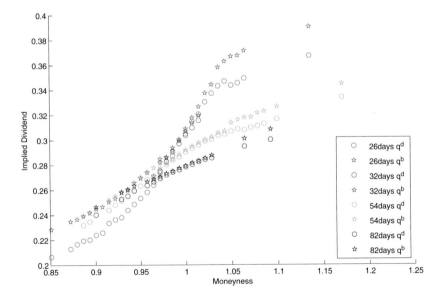

Fig. 4.7: Implied dividend curves for SPY ($\beta = 1$) options observed on November 26, 2012.

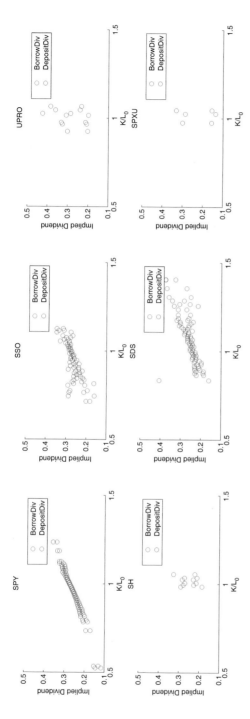

Fig. 4.8: Implied dividend as a function of adjusted moneyness for options with 1–3 months to maturity written on the (L)ETF {SPY, SSO, UPRO, SH, SDS, SPXU} with the respective leverage ratios $\{+1, +2, +3, -1, -2, -3\}$. Prices are taken from February 1, 2012 to February 28, 2012.

Figure 4.8 illustrates the implied dividends for all S&P 500 LETF options. For all LETFs, we see that the implied dividends are positive. Moreover, the two implied dividends, q^d and q^d, are closest to each other for ATM options. The spread between q^b and q^b is smallest in the non-leveraged ETF market, and largest in the triple short ($\beta = -3$) LETF market. Note that the three markets {SH, UPRO, SPXU} offer more options than shown in the figure, but many of the option prices have very wide bid-ask spreads, which are removed for this analysis. The double long/short LETFs show similar upward sloping implied dividend curves as the non-leveraged market. The implied dividend spread is wider for options with low adjusted moneyness in the SSO ($\beta = 2$) market, and for options with high adjusted moneyness in the SDS ($\beta = -2$) market.

4.3 Implied Volatility

In the following sections, we study the pricing problem for options written on LETFs and investigate the effects of leverage. We discuss a no-arbitrage valuation methodology that yields consistent prices among options on LETFs that have different leverage ratios but the same reference index. We analyze and compare prices of LETF options through their implied volatilities. In Section 4.4, we study a consistent pricing approach for LETF options under the Heston stochastic volatility model. In Section 4.5, we calibrate the model to market data and discuss the implied volatility discrepancy across leverage ratios. In Section 4.7, we incorporate random jumps in the dynamics of the reference index and propose a numerical pricing algorithm that allows for a general jump size distribution. Our algorithm is compared to existing pricing methods through some numerical examples.

Since LETF options are relatively new to the market, there is little literature on the valuation of these options. In their theses, Russell (2009) analyzes LETF implied volatilities under the CEV model, and Zhang (2010) provides numerical results for pricing LETF options assuming the reference index follows the Heston model. Ahn et al. (2013) present an approximate pricing method to compute LETF option prices with Heston stochastic volatility along with Gaussian jumps for the underlying. On the other hand, Deng et al. (2013) discuss the patterns of empirical LETF implied volatilities and compare with the simulated implied volatilities from the Heston model. Leung et al. (2014) and Leung and Sircar (2015) apply asymptotic techniques

to derive an approximation for both the LETF option price and implied
volatility under, respectively, the multiscale stochastic volatility and local-
stochastic volatility frameworks. They discuss two related concepts of *moneyness scaling* that are useful for implied volatility comparison across leverage
ratios.

To define the implied volatility of an LETF option, we first consider the
pricing problem under the Black-Scholes model. If the reference index is modeled by the geometric Brownian motion (see (2.3) with constant μ and σ),
then the risk-neutral dynamics of the LETF satisfies the Black-Scholes model,
namely,

$$\frac{dL_t}{L_t} = rdt + \beta\sigma dW^{\mathbb{Q}},$$

where $W^{\mathbb{Q}}$ is the standard Brownian motion under the risk-neutral measure $\mathbb{Q}$. For simplicity, in this chapter we assume zero expense fees for the
LETFs unless noted explicitly otherwise.

The no-arbitrage price of a European call option on L with terminal payoff
$(L_T - K)^+$ on date T is given by

$$C(t, L) := e^{-r(T-t)}\mathbb{E}^{\mathbb{Q}}\{(L_T - K)^+ \mid L_t = L\} \tag{4.8}$$

$$= C_{BS}(t, L; K, T, r, |\beta|\sigma), \quad 0 \le t \le T \tag{4.9}$$

where $C_{BS}(t, L; K, T, r, \sigma)$ is the standard Black-Scholes formula for a call
option with strike K and expiration date T when the interest rate is r and
volatility parameter is σ. We observe from equation (4.8) that theoretical
call prices on the LETF L can be expressed in terms of the traditional Black-
Scholes formula with volatility scaled by the *absolute* value of the leverage
ratio β.

One useful way to compare option prices, especially across strikes and
maturities, is to express each of them in terms of its implied volatility (IV),
where the IV is defined in the usual way:

$$\sigma_{IV}(K, T) = (C_{BS})^{-1}(C^{obs}). \tag{4.10}$$

As a consequence of (4.8), in the special case that the observed option price
coincides with the Black-Scholes model price, i.e., $C^{obs} = C_{BS}(t, L; K, T, r,$
$|\beta|\sigma)$, then it follows from that the implied volatility reduces to

$$\sigma_{IV}(K, T) = C_{BS}^{-1}(C_{BS}(t, L; K, T, r, |\beta|\sigma)) = |\beta|\sigma.$$

The same conclusion follows for put options. Therefore, ideally, we would expect that the ratio $\sigma_{IV}/|\beta|$ should be constant and equal to σ. Nevertheless, in reality one observes implied volatility skews for LETF options that do not respect this relationship. One key question is how the level of skews (slope) depends on the leverage ratio.

Since the LETF price L is again a lognormal process, the computation of all Greeks for LETF options follows in the same vein as in the Black-Scholes model. Here, we comment on the role of leverage ratio β on Delta only. The Delta represents the holding in the LETF L in the dynamic hedging portfolio and is defined by the partial derivative with respect to the LETF value L:

$$\frac{\partial C_{BS}^{(\beta)}}{\partial L}(t, L; K, T, r, \sigma) = e^{-c(T-t)}\Phi(d_1^{(\beta)}),$$

where Φ is the standard normal c.d.f. and

$$d_1^{(\beta)} = \frac{\log(L/K) + (r + \frac{\beta^2\sigma^2}{2})(T-t)}{|\beta|\sigma\sqrt{T-t}}. \tag{4.11}$$

One interesting question is how the Greeks of an ETF option is related to that of a β-LETF option since these options share the same source of randomness. We shall revisit this in Section 4.6.

Let us first draw insight from a model-free bound on the slope of the LETF option implied volatility. Holding r, T, and β fixed, with the current β-LETF price L, we denote for each β-LETF the implied volatility curve by $I^{(\beta)}(K)$, considered as a function of strike K of the β-LETF option.

The slope of the implied volatility curve admits the following bound:

$$-\frac{e^{-r(T-t)}}{|\beta|L\sqrt{T-t}}\frac{(1-\Phi(d_2^{(\beta)}))}{\Phi'(d_1^{(\beta)})} \leq \frac{\partial I^{(\beta)}(K)}{\partial K} \leq \frac{e^{-r(T-t)}}{|\beta|L\sqrt{T-t}}\frac{\Phi(d_2^{(\beta)})}{\Phi'(d_1^{(\beta)})}, \tag{4.12}$$

where d_1^{β} is given in (4.11) with $\sigma = I^{\beta}(K)$ and $d_2^{(\beta)} = d_1^{(\beta)} - |\beta|I^{(\beta)}(K)\sqrt{T-t}$.

To see this, we consider the implied volatility $I(K)$, as a function of K, and it satisfies the equality $C^{obs} = C_{BS}(t, L; K, T, r, |\beta|I^{(\beta)}(K))$. In the absence of arbitrage, the observed and Black-Scholes call option prices must be decreasing in strike K. By chain rule, we have

$$\frac{\partial C^{obs}}{\partial K} = \frac{\partial C_{BS}}{\partial K}(|\beta|I^{(\beta)}(K)) + |\beta|\frac{\partial C_{BS}}{\partial \sigma}(|\beta|I^{(\beta)}(K))\frac{\partial I^{(\beta)}(K)}{\partial K} \leq 0, \tag{4.13}$$

where the derivatives $\partial C_{BS}/\partial K$ and $\partial C_{BS}/\partial K$ are evaluated with the volatility parameter being $|\beta|I^{(\beta)}(K)$.

Since the Black-Scholes Vega $\frac{\partial C_{BS}}{\partial \sigma} > 0$ for calls and puts, we rearrange to obtain the upper bound on the slope of the implied volatility curve:

$$\frac{\partial I^{(\beta)}(K)}{\partial K} \leq -\frac{1}{|\beta|}\frac{\partial C_{BS}/\partial K}{\partial C_{BS}/\partial \sigma}(|\beta|I^{(\beta)}(K)). \tag{4.14}$$

Similarly, the fact that put prices are increasing in strike K implies the lower bound:

$$\frac{\partial I^{(\beta)}(K)}{\partial K} \geq -\frac{1}{|\beta|}\frac{\partial P_{BS}/\partial K}{\partial P_{BS}/\partial \sigma}(|\beta|I^{(\beta)}(K)), \tag{4.15}$$

where P_{BS} is the Black-Scholes put option price. Recall that puts and calls with the same K and T also have the same implied volatility $I^{(\beta)}(K)$. Hence, combining (4.14) and (4.15) gives (4.12). Implied volatility bounds like (4.12) for vanilla European options date back to Hodges (1996).

The bound (4.12) means that the slope of the implied volatility curve cannot be too negative or too positive. It also reflects that a higher leverage ratio in absolute value will scale both the upper and lower bounds, making the bounds more stringent. Intuitively, this suggests that the implied volatility curve for LETF options would be flatter than that for ETF options. Our empirical analysis also confirms this pattern over a typical range of strikes. We observe however that the leverage ratio β also appears in $d_1^{(\beta)}$ and $d_2^{(\beta)}$, so the overall scaling effect is nonlinear.

4.4 Pricing Under Heston Stochastic Volatility

We now consider a stochastic volatility model for the reference index S. Under a given risk-neutral pricing measure $\mathbb{Q}$, we assume that S follows the Heston dynamics:

$$\frac{dS_t}{S_t} = rdt + \sigma_t\,dW_t^{\mathbb{Q}}, \tag{4.16}$$

$$d\sigma_t^2 = \kappa\left(\vartheta - \sigma_t^2\right)dt + \zeta\sigma_t\,d\hat{W}_t^{\mathbb{Q}}. \tag{4.17}$$

Here, $W^{\mathbb{Q}}$ and $\hat{W}^{\mathbb{Q}}$ are two standard $\mathbb{Q}$-Brownian motions with instantaneous correlation ρ, and σ^2 is a Cox-Ingersoll-Ross (CIR) process (Cox et al. (1985)).

The corresponding leveraged ETF is constructed by a constant proportion portfolio strategy with leverage ratio β. Assuming continuous rebalancing, the price process L of the LETF is given by

$$\frac{dL_t}{L_t} = \beta \frac{dS_t}{S_t} + (1 - \beta) \, rdt \tag{4.18}$$

$$= rdt + |\beta|\sigma_t \, d\tilde{W}_t^{\mathbb{Q}}, \tag{4.19}$$

where we have denoted $\tilde{W}^{\mathbb{Q}} := \text{sign}(\beta)W^{\mathbb{Q}}$. From (4.19), we observe that the risk-neutral LETF price also satisfies the Heston model, though with different parameters. Precisely, the LETF's stochastic volatility has been scaled, $\sigma_{L,t} \equiv |\beta|\sigma_t$, and follows the re-parametrized CIR dynamics

$$d\sigma_{L,t}^2 = \kappa \left(\beta^2 \vartheta - \sigma_{L,t}^2 \right) + |\beta|\zeta\sigma_{L,t} d\hat{W}_t^{\mathbb{Q}}. \tag{4.20}$$

This also means that the instantaneous correlation between $\tilde{W}^{\mathbb{Q}}$ and $\hat{W}^{\mathbb{Q}}$ is $\rho_L = \text{sign}(\beta)\rho$. To summarize, if one assigns the reference index S to the Heston model with the set of parameters $(\sigma_0, \kappa, \theta, \zeta, \rho)$, then the β-leveraged ETF also follows the Heston model with parameters

$$\left(|\beta|\sigma_0, \kappa, \beta^2\theta, |\beta|\zeta, \text{sign}(\beta)\rho \right). \tag{4.21}$$

As a result, we see that the leverage ratio effectively magnifies the stochastic variance long-run mean θ by a factor of β^2, and the volatility level σ_0 and the volatility of volatility ζ by a factor of $|\beta|$, though the speed of mean reversion remains the same.

Remark 4.1. More generally, if the reference index S follows the dynamics

$$\frac{dS_t}{S_t} = rdt + \sigma_t dW_t^{\mathbb{Q}}, \tag{4.22}$$

with a stochastic volatility process $\{\sigma_t\}_{t\geq 0}$, then the β-LETF value follows

$$\frac{dL_t}{L_t} = rdt + \sigma_{L,t} d\tilde{W}_t^{\mathbb{Q}}, \tag{4.23}$$

where, as before, $\sigma_{L,t} = |\beta|\sigma_t$. A natural question is whether both S and L belong to the same model only with different parameters. In fact, well-known stochastic volatility models such as the Heston, Stein-Stein, Geometric, and 3/2 models satisfy this property. In Table 4.1 we summarize the ETF/LETF dynamics for this set of stochastic volatility models. A common feature of these models is that the volatility process does not explicitly depend on S, but is a function of another diffusion process. In contrast, the CEV and SABR models do not satisfy this property and S and L do not belong to the same model (see Leung and Sircar (2015)).

SV Model	Volatility Dynamics
Heston	$d\sigma_{L,t}^2 = \kappa \left(\beta^2 \vartheta - \sigma_{L,t}^2 \right) dt + |\beta| \zeta \sigma_{L,t} \, d\hat{W}_t^{\mathbb{Q}}$
Stein-Stein	$d\sigma_{L,t} = \kappa \left(|\beta| \vartheta - \sigma_{L,t} \right) dt + |\beta| \zeta \, d\hat{W}_t^{\mathbb{Q}}$
Geometric	$d\sigma_{L,t} = \kappa \sigma_{L,t} dt + \zeta \sigma_{L,t} \, d\hat{W}_t^{\mathbb{Q}}$
3/2	$d\sigma_{L,t}^2 = \kappa \left(\sigma_{L,t}^2 - \dfrac{\vartheta}{\beta^2} \sigma_{L,t}^4 \right) dt + \dfrac{\zeta}{|\beta|} \sigma_{L,t} \, d\hat{W}_t^{\mathbb{Q}}$

Table 4.1: The reference index and the LETF follow the same model dynamics for several stochastic volatility models. The dynamics of the prices are as in (4.22) and (4.23). As shown, the volatility dynamics depend on β (the reference index dynamics can be retrieved by setting $\beta = 1$).

The no-arbitrage price of a European call written on the LETF L with strike K and maturity T is given by the risk-neutral expectation:

$$C(L, \sigma) = \mathbb{E}^{\mathbb{Q}} \left\{ e^{-rT} (L_T - K)^+ \, | L_0 = L, \sigma_0 = \sigma \right\}.$$

Since the LETF follows the Heston model, whose characteristic function is known analytically, it is easy to price European options on leveraged ETFs making use of available transform methods. For example, the methods of Carr and Madan (1999), Lee (2004), and Lord et al. (2008), among others, can be used to price European options. For our computations, we first denote by

$$\Psi_L(\omega) \equiv \mathbb{E}^{\mathbb{Q}} \left\{ e^{i\omega \log \frac{L_T}{L_0}} \, | \sigma_0 = \sigma \right\}$$

the characteristic function associated with the terminal LETF price. Applying the model parameters according to (4.21) Ψ_L reads[1]

---

[1] See del Baño Rollin et al. (2009) for a summary of results on the Heston characteristic function.

$$\psi_L(\omega) = \frac{\kappa\theta}{\zeta^2}\left((\kappa - \rho\beta\zeta\omega i - d)\,t - 2\log\left(\frac{1 - ge^{-dt}}{1 - g}\right)\right)$$
$$+ \frac{\kappa - \rho\beta\zeta\omega i - d}{\zeta^2}\frac{1 - e^{-dt}}{1 - ge^{-dt}}\sigma^2, \qquad (4.24)$$

with

$$g = \frac{\kappa - \rho\beta\zeta\omega i - d}{\kappa - \rho\beta\zeta\omega i + d}, \quad \text{and} \quad d = \sqrt{(\kappa - \rho\beta\zeta\omega i)^2 + (\beta\zeta)^2\,(\omega i + \omega^2)}.$$

We can then write the option price as

$$C\left(L,\sigma\right) = \int_{-\infty}^{\infty}\left(e^{\log(L)+x} - K\right)^+ f(x)dx = \left(e^{\cdot} - K\right)^+ * f(-\cdot)\left(\log(L)\right),$$

where f denotes the probability distribution function (p.d.f.) of $\log(\frac{L_T}{L_0})$ and $*$ denotes the convolution operator. Denoting with $\mathcal{F}\left(g\left(x\right)\right) \equiv \int_{\mathbb{R}} e^{-i\omega x}g\left(x\right)dx$ the Fourier operator acting on the function g, we can then write the price of the European call as

$$C\left(L,\sigma\right) = e^{\gamma\log(L)}\mathcal{F}^{-1}\left(\mathcal{F}\left(e^{-\gamma x}\left(Le^x - K\right)^+\right)\Psi\left(\omega - i\gamma\right)\right), \quad (4.25)$$

where the introduction of the dampening factor $e^{-\gamma x}$ is necessary because the payoff is not integrable.[2] We can then apply an FFT algorithm to efficiently calculate the option price according to (4.25).

In summary, we choose to work with the Heston model because it is *leverage-invariant*, i.e., if the reference follows the Heston model, so does the LETF. There are many well-known stochastic volatility models that *not* leverage-invariant due to the path-dependence of the LETF on the reference index. Besides being more elegant, a leverage-invariant model allows us to understand each LETF market through the same set of parameters as the reference model. This would not be possible if we consider other non-leverage-invariant models, and the LETF price dynamics may not be tractable for pricing and illustration purposes. In particular, Heston has very intuitive explanation for the model parameters. This is also helpful when we explain the calibration results and price discrepancies in the next section. An additional practical benefit of the Heston model is that the tractable pricing methods and results can be reused for any LETF market, and the model can be easily calibrated with different LETF options data.

---

[2] In order for $\Psi\left(\omega - i\gamma\right)$ to be well defined, we must also ensure that $\mathbb{E}^{\mathbb{Q}}\left\{S^\gamma\right\} < \infty$.

4.5 Model Calibration and Consistency

We now discuss the procedure to calibrate the Heston model to the market prices of non-leveraged and leveraged ETF options. As is standard, given a set of market options prices C_i, $i = 1, \ldots, N$, we obtain the respective implied volatilities (IVs), $\sigma_{IV,i}$, by inverting the Black-Scholes price formula. In essence, our calibration problem involves solving the minimization problem

$$\min_{\Theta} \sum_{i=1}^{N} \omega_i |\sigma_{IV,i} - \hat{\sigma}_{IV,i}|^2 \,, \tag{4.26}$$

where $\hat{\sigma}_{IV,i}$ is the implied volatility computed based on the Heston model option price with parameters $\Theta \equiv \{\sigma_0, \kappa, \vartheta, \zeta, \rho\}$. The weights ω's will be chosen according to the liquidity of the options.

LETF options share the common source of risk due to the same reference index S. Moreover, recall that if the reference index S admits the Heston dynamics, then all the β- LETFs also follow the re-parametrized Heston model according to (4.21). This gives rise to the question of whether the parameters obtained from calibrating to one LETF options market are comparable with those from other LETF options markets. In other words, are the Heston model parameters $(\sigma_0, \kappa, \theta, \zeta, \rho)$ stable across leverage ratios? Our calibration results will shed light on this issue.

Before presenting the calibration results, let us briefly discuss the data preparation and calibration procedures. We obtain the option price data from the OptionMetrics IVY database. To enhance the calibration accuracy and efficiency, we first filter the data as follows:[3]

1. We eliminate the entries with

   - no IV values;
   - bid quotes less than \$0.5;
   - strikes for which only a call or a put is available but not both;
   - extreme moneyness, i.e., $M < 0.85$ or $M > 1.15$ for $M := K/S$.

2. Moreover, for each pair of strike and maturity (K, T), we interpolate the IVs of the associated put and call to obtain a unique value. In particular, for any given strike K, denoted by $\sigma_{IV,P}$ and $\sigma_{IV,C}$ respectively for the put and call, we set the final IV value σ_{IV} by convex combination, namely,

---

[3] Our filtering criteria are adapted from Figlewski (2010) and Chapter 5.3 of Fouque et al. (2011).

$$\sigma_{IV} = w\sigma_{IV,P} + (1-w)\sigma_{IV,C}, \tag{4.27}$$

with the assigned weight $w := (1.15S - K)/(0.3S)$. In other words, the parameter w weights the implied volatility of out-of-the-money options more.

In addition, we also weigh the IVs according to the liquidity of the options. Specifically, we choose ω_i to be inversely proportional to the (relative) bid-ask spread,

$$\omega_i = c\,\frac{b_i + a_i}{2|a_i - b_i|}, \tag{4.28}$$

where b_i and a_i denote the bid and ask prices of option i, respectively, and

$$c = \frac{N}{\sum_{i=1}^{N} \frac{b_i+a_i}{2|a_i-b_i|}}$$

is a normalizing constant so that $\sum_{i=1}^{N} \omega_i = N$. As a consequence of our choice, less liquid options will be considered less relevant during the calibration procedure.

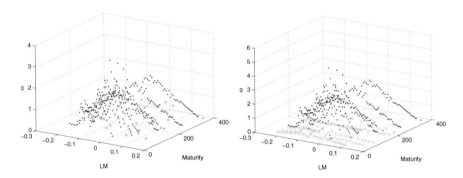

Fig. 4.9: The weights, ω, obtained from the bid-ask spreads on the SPY (dots) and UPRO (crosses) options on June 25, 2013. In the top panel the normalizing constant in (4.28) is computed for the SPY and URPO separately. In the bottom panel the normalizing constant is the same, allowing a direct comparison of liquidity between SPY and UPRO options.

In Figure 4.9, we plot the ω weights for the options written, respectively, on SPY and UPRO on June 25, 2013. In the top panel we show the weights for each ETF separately, i.e., where the normalizing constant c is calculated for the two set of options. Therefore, the proportionality to the bid-ask spread is valid only between omega values of options on the same ETF and not between options on different ETFs. This picture illustrates the different liquidity across strikes and maturities for each ETF market. As we can see, the weights are higher for options close to the ATM strike for both ETFs, and for shorter maturities as well. In the bottom panel, we calculate the ω weights for the same ETFs but where the normalizing constant is calculated on the union of the two sets. This way, it is possible to compare the liquidity of options on the two ETFs and these weights could be used to calibrate a model on the collection of options on both ETFs. It is not surprising then that the ω weights for UPRO appear to be much lower than those for SPY due to the presence of wider bid-ask spreads on the less liquid UPRO options. For the same reason then, when calibrating a model on the entire collection of data on both ETFs one might want to use the ω weights showed in the top panel. Otherwise, because of the discrepancy in liquidity, the results would be affected only by the data of the most liquid ETF.

In Figures 4.10(a) and 4.10(b), we show the observed, cleaned IVs for the selected (L)ETFs on the S&P 500 index on Tuesday, June 25, 2013. For the SPY ETF, we have the following maturities, listed as calendar days to expiration: 24, 52, 87, 115, 178, 206, 269, 360. For other LETFs (besides SH), we have a subset of these maturity dates. We calibrate the Heston model repetitively by solving problem (4.26) using options on each LETF with a different leverage ratio. Notice that for each LETF the ω weights in (4.26) are calculated according to (4.28) using options on that particular ETF. To price the options we adopt the approach introduced in Section 4.4 and to perform the calibration we adapt a trust-region-reflective gradient-descent algorithm (Coleman and Li (1994); Coleman and Li (1996)), starting from different initial points to guarantee a better exploration of the parameters space. While our numerical tests show the adopted method results in an effective calibration, we remark that there are many alternative, possibly more advanced, calibration procedures available (see, e.g., Cont and Tankov (2002) and references therein).

The results of the calibration are reported in Table 4.2 and Figure 4.11. Notice that for each LETF, the Heston model parameters reported in Table 4.2 are with respect to the reference, non-leveraged index consistently with the transformation (4.21). This way the models are directly comparable to each other. The mean relative cross calibration error, $\epsilon_{j,k}$, is calculated separately for each pair of LETFs (denoted by the tuple (j,k)), and is defined as the arithmetic mean of the relative error computed on each option on LETF j when using the calibration parameters obtained for LETF k:

$$\epsilon_{j,k} \equiv \frac{1}{N_j} \sum_{i=1}^{N_j} \frac{|\sigma_{IV,i} - \hat{\sigma}_{IV,i}(\Theta_k)|}{\sigma_{IV,i}}, \qquad (4.29)$$

where the sum is over the N_j options available for LETF j, and we have denoted by $\hat{\sigma}_{IV,i}(\Theta_k)$ the model IV obtained for option i of LETF j when applying the parameters obtained through calibration for LETF k, Θ_k.

Although the parameters do not exactly match, they are often comparable to each other. In addition, we notice that the greatest error difference is between long and short LETFs, particularly with respect to the parameter κ. For example, as we can see from Figure 4.11, when we calibrate the model using SPY options data, we obtain an error of around 1% on the same SPY options, and of about 2% for SSO and UPRO, while the error is around 3.5% and 4% for SDS and SPXU, respectively. Overall, we can see that, for each line, the points relative to the long LETFs are closer to each other than to those relative to the short LETFs, and vice versa.

ETF	β	σ_0^2	κ	θ	ρ	ζ
SPY	1	0.032	3.1	0.052	-0.75	0.89
SSO	2	0.031	2.2	0.056	-0.77	0.85
UPRO	3	0.031	2.3	0.057	-0.84	0.82
SH	-1	0.033	4.4	0.035	-0.93	0.55
SDS	-2	0.032	8.9	0.036	-0.84	0.93
SPXU	-3	0.033	10	0.033	-0.85	0.85

Table 4.2: Parameters obtained by approximately solving (4.26) under the Heston model for each LETF on data obtained for Tuesday, June 25, 2013. See Figures 4.10(a), 4.10(b) for the observed IVs.

LETF	σ_0^2	κ	θ	ρ	ζ
Long	0.031	2.7	0.054	-0.79	0.86
Short	0.033	9.5	0.035	-0.79	1.01

Table 4.3: The calibrated parameters when solving (4.26) for the aggregate long ($\beta > 0$) and short ($\beta < 0$) LETF IVs on June, 25, 2013. Compare with Table 4.2.

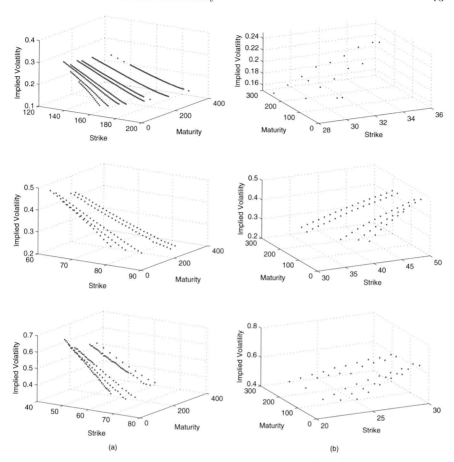

Fig. 4.10: The observed IVs prepared for the selected LETFs on Tuesday, June 25, 2013. In panel (a), from top to bottom, we have the SPY, SSO, and UPRO, with $\beta = +1, +2, +3$ respectively, IV surfaces. In panel (b), from top to bottom, we have the SH, SDS, and SPXU, with $\beta = -1, -2, -3$ respectively, IV surfaces.

Motivated by the observations above, we then perform two separate calibrations, one on long and one on short LETFs. The results of these calibrations are listed in Table 4.3 and Figures 4.12, 4.13, and 4.14. In particular, in each plot of Figure 4.13 we superimpose the calibrated IVs obtained by solving problem (4.26) on the long LETFs options on the observed IVs. Similarly, in Figure 4.14 we show those obtained by calibrating on the short LETFs options. Once again, the results show the difference between the two

classes of LETFs. A calibration that aims to reduce the error on one class
will produce a somewhat higher error on the other. From Figure 4.13 we can
observe that the model calibrated on the long LETFs data seem to overprice
options with positive moneyness on the short LETFs while slightly under-
pricing those with negative moneyness. Interestingly, the same observation
holds when we reverse the role of the long and short LETFs, as exemplified
in Figure 4.14.

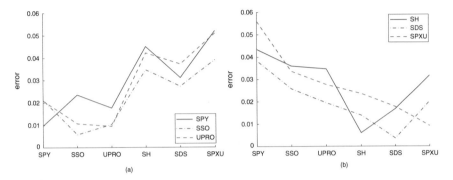

Fig. 4.11: The cross calibration errors calculated according to (4.29). The
model parameters are taken from Table 4.2. Panels (a) and (b) report the
cross calibration errors using the parameters obtained, respectively, for the
long and short LETFs.

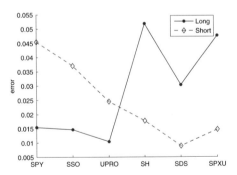

Fig. 4.12: The cross calibration errors computed according to (4.29) when the
model parameters are as in Table 4.3 and the calibration has been performed
on aggregate long ($\beta > 0$) and short ($\beta < 0$) LETF option data.

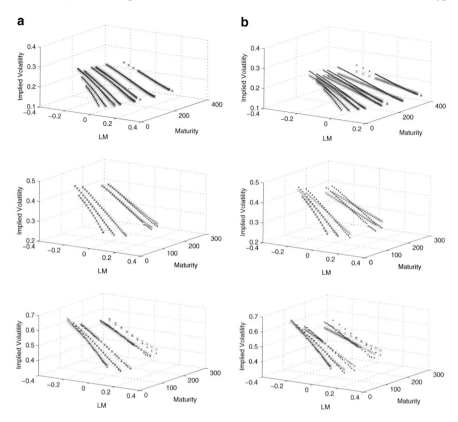

Fig. 4.13: The IV surfaces for short LETFs when the model parameters are calibrated, respectively, from long (panel (a)) and short (panel (b)) LETF options data; see Table 4.3. The dotted points represent observed IVs while the crossed ones represent calibrated IVs.

4.6 Moneyness Scaling

We now introduce the idea of linking implied volatilities between ETF and LETF options via the method of *moneyness scaling*. This will be useful for traders to compare the option prices across not only different strikes but also different leverage ratios, and potentially identify option price discrepancies. One plausible explanation for the different IV patterns is that different leverage ratios correspond to different underlying dynamics. Therefore, the distribution of the terminal price of any β-LETF naturally depends on the leverage ratio β.

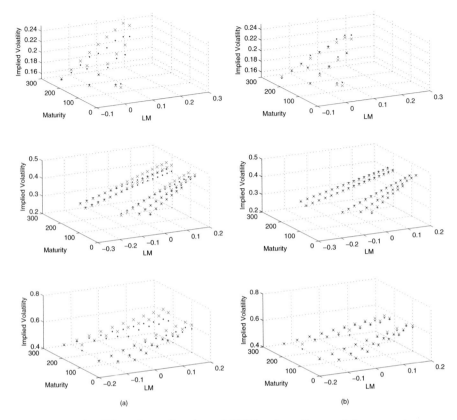

Fig. 4.14: The IV surfaces for short LETFs when the model parameters are calibrated, respectively, from short (panel (a)) and long (panel (b)) LETF options data; see Table 4.3. The dotted points represent observed IVs while the crossed ones represent calibrated IVs.

In a general stochastic volatility model, we can write the log LETF price as

$$\log\left(\frac{L_T}{L_0}\right) = \beta \log\left(\frac{S_T}{S_0}\right) - (r(\beta-1)+f)T - \frac{\beta(\beta-1)}{2}\int_0^T \sigma_t^2\, dt,$$

where $(\sigma_t)_{t\geq 0}$ is the stochastic volatility process of the reference index S. The terms $\log\left(\frac{L_T}{L_0}\right)$ and $\log\left(\frac{S_T}{S_0}\right)$ are, respectively, the log-moneyness of the terminal LETF value and terminal reference index (or non-leveraged ETF) value. As discussed in Leung and Sircar (2015), if we condition on that the terminal random log-moneyness $\log\left(\frac{S_T}{S_0}\right)$ equal to some constant $LM^{(1)}$, then the conditional expectation of the β-LETF log-moneyness is given by

$$LM^{(\beta)} = \beta LM^{(1)} - (r(\beta - 1) + f)T$$
$$- \frac{\beta(\beta - 1)}{2} \mathbb{E}^{\mathbb{Q}} \left\{ \int_0^T \sigma_t^2 dt \,\Big|\, \log\left(\frac{S_T}{S_0}\right) = LM^{(1)} \right\}. \qquad (4.30)$$

In other words, $LM^{(\beta)}$ can be viewed as the best estimate for the terminal log-moneyness of L_T, given the terminal value $\log\left(\frac{S_T}{S_0}\right) = LM^{(1)}$. This equation strongly suggests a relationship between the log-moneyness between ETF and LETF options, though the conditional expectation in (4.30) is nontrivial. Zhang (2010) calls the LETF option strike corresponding to (4.30) the "most likely" strike.

In order to obtain a more explicit and useful relationship, let us consider the simple case with a constant σ as in the Black-Scholes model. Then, equation (4.30) reduces to

$$LM^{(\beta)} = \beta\, LM^{(1)} - (r(\beta - 1) + f)T - \frac{\beta(\beta - 1)}{2}\sigma^2 T. \qquad (4.31)$$

By this formula, the β-LETF log-moneyness $LM^{(\beta)}$ is expressed as an affine function of the non-leveraged ETF log-moneyness $LM^{(1)}$. The leverage ratio β not only scales the log-moneyness $LM^{(1)}$, but also plays a role in two constant terms. Leung et al. (2014) derive the asymptotic expansion of the LETF IVs for a wide class of local stochastic volatility models, including the CEV, SABR, and Heston models. They also related the LETF IV to the non-leveraged ETF IV and obtain a linear log-moneyness scaling like (4.31) in the first-order approximation term.

Interestingly, the moneyness scaling formula (4.31) can also be derived using the idea of *dual Delta matching* under the Black-Scholes model. To this end, the dual Deltas of the LETF call and put options are defined, respectively, by the partial derivatives of their prices with respect to strike:

$$\frac{\partial C_{BS}^{(\beta)}}{\partial K}(t, L; K, T, r, f, \sigma) = -e^{-r(T-t)}\Phi(d_2^{(\beta)}),$$
$$\frac{\partial P_{BS}^{(\beta)}}{\partial K}(t, L; K, T, r, f, \sigma) = e^{-r(T-t)}\Phi(-d_2^{(\beta)}), \qquad (4.32)$$

where

$$d_2^{(\beta)} = \frac{\log(L/K) + (r - f - \frac{\beta^2 \sigma^2}{2})(T - t)}{|\beta|\sigma\sqrt{T - t}}. \qquad (4.33)$$

Under the Black-Scholes model, the dual Delta of a β-LETF call (resp. put), with log-moneyness $LM^{(\beta)}$ defined in (4.31), coincides with

(i) the dual Delta of an ETF call (resp. put) with log-moneyness $LM^{(1)}$ if $\beta > 0$, or
(ii) the negative of the dual Delta of an ETF put (resp. call) with log-moneyness $LM^{(1)}$ if $\beta < 0$.

To see the equivalence, let us apply (4.31) to the dual Delta of the β-LETF call and get

$$
\begin{aligned}
-e^{-r(T-t)}\Phi(d_2^{(\beta)}) &= -e^{-r(T-t)}\Phi\left(\frac{-LM^{(\beta)} + (r - f - \frac{\beta^2\sigma^2}{2})(T-t)}{|\beta|\sigma\sqrt{T-t}}\right) \\
&= -e^{-r(T-t)}\Phi\left(\frac{\beta}{|\beta|} \cdot \frac{-LM^{(1)} + (r - \frac{\sigma^2}{2})(T-t)}{\sigma\sqrt{T-t}}\right) \\
&= -e^{-r(T-t)}\Phi\left(\text{sign}(\beta)\, d_2^{(1)}\right), \qquad (4.34)
\end{aligned}
$$

where $d_2^{(1)}$ is given in (4.33).

Notice that, for $\beta > 0$, equality (4.34) indeed yields the dual delta for the ETF *call* with log-moneyness $LM^{(1)}$. For $\beta < 0$, the term $-e^{-r(T-t)}\Phi(-d_2^{(1)})$ in (4.34) is the negative of the dual delta of an ETF ($\beta = 1$) *put* with log-moneyness $LM^{(1)}$ (see (4.32)).

Additionally, we can apply (4.31) to link the log-moneyness pair $(LM^{(\beta)}, LM^{(\hat\beta)})$ of LETFs with different leverage ratios $(\beta, \hat\beta)$, namely,

$$
\begin{aligned}
LM^{(\beta)} = \frac{\beta}{\hat\beta}\left(LM^{(\hat\beta)} + (r(\hat\beta - 1) + \hat f)T + \frac{\hat\beta(\hat\beta - 1)}{2}\sigma^2 T\right) - (r(\beta - 1) + f)T \\
- \frac{\beta(\beta - 1)}{2}\sigma^2 T, \qquad (4.35)
\end{aligned}
$$

where f and $\hat f$ are the expense fees associated with the β-LETF and $\hat\beta$-LETF, respectively.

In summary, the proposed moneyness scaling method is a simple and intuitive way to identify the appropriate pairing of ETF and β-LETF IVs. In other words, for each β-LETF IV, the associated log-moneyness $LM^{(\beta)}$ can be viewed as a function of the log-moneyness $LM^{(1)}$ associated with the non-leveraged ETF. Consequently, we can plot the IVs for both the β-LETF and ETF over the same axis of log-moneyness $LM^{(1)}$ (see Figure 4.15). As a result, we can compare the LETF IVs with the ETF IVs by plotting them

over the same log-moneyness axis. In practical calibration, we do not observe or assume a constant volatility σ. In order to apply moneyness scaling
formula (4.31), we replace σ with the average IV across all available strikes
for the ETF options. A more cumbersome alternative way is to estimate the
conditional expectation of the integrated variance in (4.30). The expected
variance σ_t^2 at time $t \leq T$ conditioned on the *terminal* value of the underlying is not easily computable even for specific models. It becomes explicit in
the Black-Scholes model with constant volatility, as stated in (4.31).

Next, we apply the log-moneyness scaling to the empirical IVs. In Figure 4.15, we plot both the ETF and LETF IVs after moneyness scaling on
August 23, 2010 with 54 days to maturity. As we can see, the mapping works
remarkably well as the LETF and ETF IVs overlap significantly after scaling
(right panel). In particular, the scaled IVs (right panel) for the double short
LETF, SDS, are now *downward* sloping even though the original ones (left
panel) are upward sloping. As a result, the log-moneyness scaling allows us
to visibly discern the potential mismatch between ETF and LETF IVs. It is
also useful for the market making of LETF options, especially when there is
less liquidity and fewer contracts available for a certain leverage ratio as we
typically observe in the triple short ($\beta = -3$) LETF options market. Since
the non-leveraged ETF market tends to have more strikes available, for every ETF option traded with log-moneyness $LM^{(1)}$, there may be no LETF
option traded with log-moneyness $LM^{(\beta)}$. Nevertheless, one can apply the
log-moneyness scaling to determine the implied volatility, and thus the price,
at the log-moneyness $LM^{(\beta)}$ of the currently unavailable LETF option. To
understand this, let us look at Figure 4.15 (right panel). After moneyness
scaling, we see that there is an SPY option traded at/near the moneyness
$LM = -0.4$, but not for SSO and SDS. If the market marker wants to offer
such an option on SSO or SDS, then he/she can set the price according to
the implied volatility curve of the SPY (crosses) at $LM = -0.4$, and sub-
sequently recover the LETF log-moneyness $LM^{(\beta)}$ via the scaling formula
(4.35).

4.7 Incorporating Jumps with Stochastic Volatility

We now extend the dynamics (4.16) by incorporating jumps in the reference
index. In particular, we assume that the reference ETF S follows

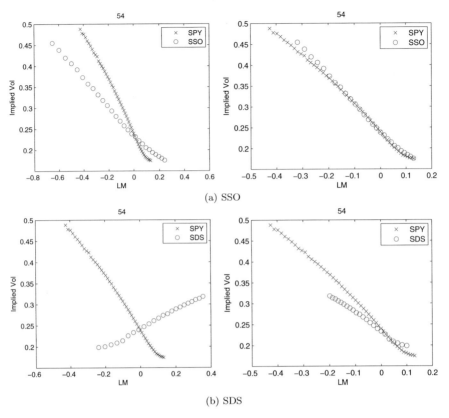

(a) SSO

(b) SDS

Fig. 4.15: SPY (crosses) and LETF (circles) implied volatilities before (left panel) and after (right panel) moneyness scaling on August 23, 2010 with 54 days to maturity, plotted against log-moneyness of SPY options.

$$\log \frac{S_t}{S_0} = \mu t + X_t + \sum_{i=1}^{N_t} Z_i, \tag{4.36}$$

$$X_t = \int_0^t \sigma_s dW_s^{\mathbb{Q}} - \int_0^t \frac{\sigma_s^2}{2} ds,$$

$$d\sigma_s^2 = \kappa \left(\vartheta - \sigma_s^2 \right) ds + \zeta \sigma_s d\hat{W}_s^{\mathbb{Q}},$$

where $W^{\mathbb{Q}}$, $\hat{W}^{\mathbb{Q}}$ are two $\mathbb{Q}$-Brownian motions with instantaneous correlation $\rho \in (-1, 1)$. N is a Poisson process with intensity λ. The random jump sizes $(Z_i)_{i=1,2,\ldots}$ are independent, identically distributed random variables whose distribution satisfies $\mathbb{E}^{\mathbb{Q}} \left\{ e^Z \right\} < \infty$. In addition, N and the jumps $(Z_i)_{i=1,2,\ldots}$ are assumed to be independent of each other and of the processes

$W^{\mathbb{Q}}$ and $\hat{W}^{\mathbb{Q}}$. Finally, we set $\mu = rt - \lambda mt$, where $m \equiv \mathbb{E}^{\mathbb{Q}}\left\{e^{Z} - 1\right\}$, so that the discounted reference price $(e^{-rt}S_t)_{t \geq 0}$ is a $\mathbb{Q}$-martingale. The stochastic volatility jump-diffusion model is related to those in the literature, including the SVJ (Bates (1996)), Merton (1976), Kou (2002), Variance Gamma (Madan and Unal (1998)), and CGMY (Carr et al. (2002)) models.

Recall that the LETF is designed to yield a multiple of the daily returns of the underlying. In principle, it is possible for the fund to experience a loss greater than 100%. However, protected by the principle of limited liability, the LETF value can never be negative. In practice, some LETF providers design the fund so that the daily returns are capped both downward and upward.[4] If we assume a continuous rebalancing frequency and LETF returns capped from below and above at the levels l and h, respectively, then LETF value follows

$$L_t = L_0 e^{(\mu_L t + X_{L,t})} \prod_{i=1}^{N_t} (1 + Y_{L,i}), \qquad (4.37)$$

$$X_{L,t} = \int_0^t \beta \sigma_s dW_s^{\mathbb{Q}} - \int_0^t \frac{\beta^2 \sigma_s^2}{2} ds,$$

$$Y_{L,i} = \min\left\{\max\left\{\beta\left(e^{Z_i} - 1\right), l\right\}, h\right\}, \qquad (4.38)$$

with $l \geq -1$, $h \in [0, +\infty]$, and

$$\mu_L \equiv r - \lambda m_L, \quad \text{and} \quad m_L \equiv \mathbb{E}^{\mathbb{Q}}\left\{Y_L\right\}.$$

As in Section 4.4, the continuous part, $(e^{X_{L,t}})_{t \geq 0}$ follows the Heston process with stochastic volatility $\sigma_{L,t} \equiv |\beta|\sigma_t$ (see equation (4.20)). On the other hand, the jump distribution of $e^Z - 1$ will generally differ from that of Y_L. This might thus introduce another layer of complexity in the pricing of options written on L.

---

[4] For example, according to the summary prospectus of the Direxion Daily S&P 500 Bull 3x Shares: "Gain Limitation Risk: If the Fund's underlying index moves more than 33% on a given trading day in a direction adverse to the Fund, you would lose all of your money. Rafferty will attempt to position the Fund's portfolio to ensure that the Fund does not lose more than 90% of its NAV on a given day. The cost of such downside protection will be limitations on the Fund's gains. As a consequence, the Fund's portfolio may not be responsive to Index gains beyond 30% in a given day. For example, if the Index were to gain 35%, the Fund might be limited to a daily gain of 90% rather than 105%, which is 300% of the Index gain of 35%." This suggests a two-sided cap on the jump sizes. Source: http://www.direxioninvestments.com/products/direxion-daily-sp-500-bull-3x-etf.

Again, we are interested in pricing the European call option given by the risk-neutral expectation

$$C\left(L,\sigma\right) = \mathbb{E}^{\mathbb{Q}}\left\{e^{-rT}\left(L_T - K\right)^+ | L_0 = L, \sigma_0 = \sigma\right\}.$$

In order to use the results of Section 4.4 and formula (4.25) to price the option, we need to calculate

$$\Psi_L(\omega) \equiv \mathbb{E}^{\mathbb{Q}}\left\{e^{i\omega\log\frac{L_T}{L_0}} | \sigma_0 = \sigma\right\}.$$

However, $\log\left(L_T\right)$ is not well defined in this case because the price L_T can actually reach zero with positive probability. To overcome this, we write

$$C\left(L,\sigma\right) = p\mathbb{E}^{\mathbb{Q}}\left\{e^{-rT}\left(L_T - K\right)^+ | L_T > 0, L_0 = L, \sigma_0 = \sigma\right\}, \qquad (4.39)$$

where

$$p \equiv \mathbb{Q}\left\{L_T > 0 | L_0 = L\right\} = e^{-\lambda T\mathbb{Q}\{Y_L \leq -1\}}. \qquad (4.40)$$

As a result, we can price the call option using (4.25) if we are able to obtain

$$\tilde{\Psi}_L(\omega) \equiv \mathbb{E}^{\mathbb{Q}}\left\{e^{i\omega\log\frac{L_T}{L_0}} | L_T > 0, \sigma_0 = \sigma\right\}. \qquad (4.41)$$

Notice that, in order to use the Fourier transform method (4.25), we also need to verify that $\mathbb{E}\left\{\log\left(L_T\right) | L_T > 0\right\} < \infty$. We analyze the validity of this condition at the end of this section.

Next, we show how to compute $\tilde{\Psi}_L$. We start by observing that the characteristic function of $\log\frac{S_T}{S_0}$ is given by

$$\mathbb{E}^{\mathbb{Q}}\left\{e^{i\omega\log\frac{S_T}{S_0}} | \sigma_0 = \sigma\right\} = e^{i\omega\mu T + \psi_X(\omega) + \lambda T\left(e^{\psi_Z(\omega)} - 1\right)}, \qquad (4.42)$$

where

$$\psi_X(\omega) \equiv \log\mathbb{E}^{\mathbb{Q}}\left\{e^{i\omega X_T} | \sigma_0 = \sigma\right\}, \quad \text{and} \quad \psi_Z(\omega) \equiv \log\mathbb{E}^{\mathbb{Q}}\left\{e^{i\omega Z}\right\}.$$

In particular, $\psi_X(\omega)$ is the characteristic exponent of the Heston model, see equation (4.24). On the other hand, the characteristic exponent of Z depends on the particular choice of the jump distribution. For example, when Z is distributed as a double-exponential random variable, $Z \sim DE(u, \eta_1, \eta_2)$, the characteristic exponent will be

$$\psi_Z(\omega) = u\frac{\eta_1}{\eta_1 - i\omega} + (1 - u)\frac{\eta_2}{\eta_2 + i\omega}.$$

We notice the event $\{L_T > 0\} = \{Y_{L,i} > -1, i = 1, \ldots, N_T\}$. Therefore, we express (4.41) as

$$\tilde{\Psi}_L(\omega) = \mathbb{E}^{\mathbb{Q}}\left\{ e^{i\omega\left(\mu_L T + X_{L,T} + \sum_{i=1}^{N_T} Z_{L,i}\right)} | L_T > 0, \sigma_0 = \sigma \right\}, \qquad (4.43)$$

where

$$Z_{L,i} \equiv \begin{cases} \log\left(1 + Y_{L,i}\right), & Y_{L,i} > -1, \\ 0, & Y_{L,i} = -1. \end{cases} \qquad (4.44)$$

Therefore, $\tilde{\Psi}_L$ is given by

$$\tilde{\Psi}_L(\omega) = e^{i\omega\mu_L T + \psi_{X_L}(\omega) + \lambda T\left(e^{\psi_{Z_L}(\omega)} - 1\right)}, \qquad (4.45)$$

where

$$\psi_{X_L}(\omega) \equiv \log \mathbb{E}^{\mathbb{Q}}\left\{ e^{i\omega X_{L,T}} | \sigma_0 = \sigma \right\},$$
$$\psi_{Z_L}(\omega) \equiv \log \mathbb{E}^{\mathbb{Q}}\left\{ e^{i\omega Z_L} | Y_{L,i} > -1 \right\}.$$

We have derived the analytic expression for $\psi_{X_L}(\omega)$ in Section 4.4, see equation (4.24). However, an explicit expression for ψ_{Z_L} is not easily obtained. Ahn et al. (2013), who use a different transform method, analyze the special case when Z is Gaussian, and circumvent the aforementioned issue by finding an analytic, approximate expression for ψ_{Z_L}. In contrast, because of the choice of the pricing formula (4.25), we are able to calculate the values of ψ_{Z_L} numerically as shown below, introducing no approximation error and providing a method that can be used for virtually any distribution of the jump Z.

Recall that Z_L is a function of Z, see (4.44) and (4.37). Therefore, if the p.d.f. of Z is known, we can easily obtain the analytic expression for the p.d.f. of Z_L. The desired values of the characteristic function $e^{\psi_{Z_L}}$ can then be easily computed via an FFT algorithm and plugged into (4.45) in order to obtain Ψ_L and price the option according to (4.25). If the p.d.f. of Z is not available but its characteristic function is, one further step can be added in order to first obtain the p.d.f. from its characteristic function via FFT. Importantly, in this case, because of the transformation (4.38) we would need to make use of a nonuniform FFT algorithm. The numerical results for this method are presented in the next section.

Finally, we would like to have that, given $L_T > 0$, the terminal log-price $\log(L_T)$ is L_1-integrable. We have showed with (4.45) that $\tilde{\Psi}_L$ can be obtained from the characteristic functions of X_L and Z_L. The results on the log-spot under the Heston model are well known (see, for example, del Baño Rollin et al. (2009)) and we do not discuss them. On the other hand, under the assumption that Z is integrable, we show that Z_L is integrable, too. For simplicity, we assume that Z admits the p.d.f. $f_Z(x)$, and that the returns of the leveraged ETF L are not capped, i.e., $l = -1$ and $h = \infty$. Given $Y_{L,i} > -1$, we can then write

$$Z_L = \log\left(\beta\left(e^{Z_i} - 1\right) + 1\right).$$

As a result, the p.d.f. of $Z_L | Y_{L,i} > -1$ can be written as

$$f_{Z_L | Y_{L,i} > -1}(x) = c\frac{\mathrm{sign}(\beta)e^x}{e^x + \beta - 1}f_Z\left(g^{-1}(x)\right), \qquad (4.46)$$

for $x \in \mathbb{R}$ if $\beta > 0$ $x \in (-\infty, \log(1 - \beta))$ if $\beta < 0$, where

$$g(x) \equiv \log\left(\beta(e^x - 1) + 1\right), \quad \text{and} \quad g^{-1}(x) = \log\left(\frac{e^x + \beta - 1}{\beta}\right),$$

along with the normalizing constant c. Therefore, we see that for very small values of x, the p.d.f. is approximately

$$f_{Z_L}(x) \approx c\frac{\mathrm{sign}(\beta)e^x}{\beta - 1}f_Z\left(\log(\frac{\beta - 1}{\beta})\right), \quad x \ll 0, \qquad (4.47)$$

while, for very big values of x (which is possible only when $\beta > 0$),

$$f_{Z_L}(x) \approx cf_Z\left(x - \log\beta\right), \qquad x \gg 0. \qquad (4.48)$$

Therefore, the integrability of Z is sufficient to guarantee the integrability of Z_L.

In Table 4.4, we present the LETF option prices computed via formula (4.25) and via a standard Monte Carlo algorithm with an Euler scheme applied to the SVJ process. For comparison, in the same table, we also report the prices[5] obtained by Ahn et al. (2013)(AHJ) for the same case and we discuss the differences below.

---

[5] See pages 14–18 of their paper.

As we can observe from Table 4.4, the error between our prices obtained via formula (4.25) and via the Monte Carlo method are in good agreement. Beside the case of negatively leveraged ETF, the error is virtually zero. On the other hand, we notice that AHJ prices, which are also calculated via a transform and Monte Carlo method, differ from ours. We stand to explain the differences as follows:

- In their Monte Carlo implementation, AHJ opt to simulate a process where the LETF is rebalanced daily while the dynamics and transform method adopted assume continuous rebalancing.
- In their Fourier transform implementation, AHJ approximate the analytic form of the characteristic function Ψ_L and use this when applying the Carr and Madan (1999) algorithm to obtain prices. Instead, we calculate the characteristic function of Ψ_L numerically, without introducing further approximations, and easily deploy it, thanks to the use of a different formula for pricing, based on the transform of a convolution, namely expression (4.25).

In Table 4.4, the pricing errors from both methods are very small and therefore negligible for practical applications. Nevertheless, our method has some advantages because 1) it does not introduce any approximation and is in principle more accurate, with the only error in pricing being attributed to the discretization of the Fourier integral; and 2) it is more general because it can accommodate a variety of models and jump distributions for the underlying.

Next, we discuss other features of the model (4.37) by analyzing numerical results obtained through the implementation of (4.45). In Figure 4.16, we show the IV surfaces for a particular instance of the SVJ model when $\rho \neq 0$. It is evident that, in this model, prices for long and short LETFs with the same absolute leverage ratio are not equal. In contrast, under the BS framework with continuous rebalancing, prices of similar options on LETFs with opposite leverage ratios are equal. This is a result of the fact that the LETFs dynamics under BS admit marginal distributions that are symmetric in the leverage parameter β. However, this property does not hold for many other models. In particular, under the Heston model, marginals are not symmetric because the parameter ρ_L satisfies $\rho_L(\beta) = -\rho_L(-\beta)$. In fact, all other model parameters are symmetric in β and prices are symmetric when $\rho = \rho_L = 0$. In particular, this means that, while for a long LETF out-of-the-money (OTM) puts have higher IV than OTM calls, for a short LETF the contrary is true. This is rather intuitive if one thinks that an OTM put on a long LETF is a bet on the price of the underlying going downward, similar to the bet expressed by an OTM call on the short LETF.

β	M_S	M_L	SV				SVJ				error	
			FT_{AHJ}	MC_{AHJ}	FT_{LS}	MC_{LS}	FT_{AHJ}	MC_{AHJ}	FT_{LS}	MC_{LS}	FT_{AHJ}	FT_{LS}
1	0.9	0.9	33.66	33.66	33.66	33.66	33.66	33.66	33.66	33.66	0.00	0.00
1	1	1	20.10	20.1	20.10	20.10	20.05	20.05	20.05	20.05	0.00	0.00
1	1.1	1.1	11.23	11.23	11.23	11.23	11.22	11.23	11.22	11.22	0.00	0.00
2	0.75	0.5	60.66	60.74	60.66	60.66	61.41	61.51	61.44	61.44	-0.03	0.00
2	1	1	37.78	37.87	37.78	37.78	38.41	38.43	38.37	38.37	0.04	0.00
2	1.25	1.5	24.11	24.18	24.11	24.11	24.52	24.45	24.41	24.41	0.11	0.00
3	0.75	0.25	81.82	81.98	81.82	81.82	83.08	83.25	83.07	83.07	0.01	0.00
3	1	1	52.77	53.09	52.77	52.77	53.98	54.07	53.87	53.87	0.11	0.00
3	1.25	1.75	37.30	37.6	37.30	37.30	37.91	37.84	37.69	37.69	0.22	0.00
-1	0.75	1.25	14.14	14.15	14.14	14.14	13.87	13.93	13.90	13.90	-0.03	0.00
-1	1	1	21.15	21.19	21.15	21.15	21.03	21.16	21.10	21.10	-0.07	0.00
-1	1.25	0.75	32.73	32.79	32.72	32.72	33.01	33.24	33.16	33.16	-0.15	0.00
-2	0.75	1.5	31.88	32.09	31.88	31.88	30.92	31.29	31.01	31.02	-0.10	-0.01
-2	1	1	41.68	41.95	41.68	41.68	41.13	41.6	41.26	41.27	-0.14	-0.01
-2	1.25	0.5	60.02	60.25	60.02	60.02	60.56	61.09	60.76	60.77	-0.21	-0.01
-3	0.75	1.75	50.72	51.48	50.72	50.72	48.47	49.43	48.46	48.74	-0.27	-0.28
-3	1	1	60.17	60.88	60.17	60.17	58.57	59.56	58.59	58.87	-0.30	-0.28
-3	1.25	0.25	81.46	81.74	81.46	81.46	81.82	82.59	81.89	82.17	-0.35	-0.29

Table 4.4: LETF option prices calculated according to model (4.37) with a Gaussian jump, $Z \sim N(\mu_Z, \sigma_Z)$. Our prices are calculated according to a simple Euler MC scheme (MC_{LS}) and the Fourier transform (FT_{LS}) method presented in Section 4.4. For comparison, the prices from Ahn et al. (2013) (pg.18) are also reported. The errors are calculated as the difference between the respective Fourier transform prices and our MC prices. Parameters: $r = .01, \kappa = .65, \zeta = .7895, \theta = .3969, \rho = .3969, \sigma_0^2 = .7571, \lambda = 2.1895, \mu = .0105, \sigma_Z = .2791$.

Figure 4.17 illustrates how the IV smiles are affected by the choice of different caps and floors for the LETF returns. In both the left and right plots, we observe that prices are monotonically increasing with the absolute value of the threshold l or h. This is somewhat intuitive if one realizes that capping returns lowers the variance of the stock price which, altogether with the requirement that the stock price be a martingale, reduces the averaged payoff and option price.

Finally, in Figure 4.18 we show the jump distribution Z_L under the SVJ model for different values of β and σ when $l = -1$ and $h = \infty$. We notice that the leveraged ETFs have a fatter left tail and a thinner right tail, when compared to the non-leveraged underlying. This observation can be quantified when we recall the form of the p.d.f. of Z_L obtained in (4.46). As shown in

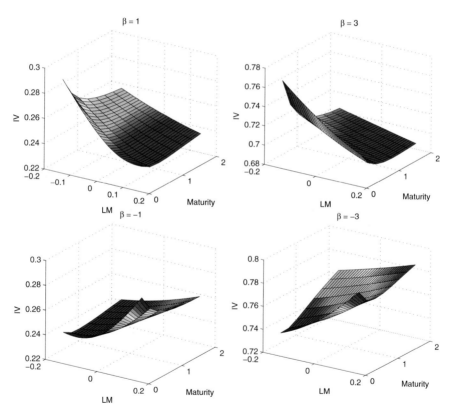

Fig. 4.16: The IV surfaces against log-moneyness (LM) and maturity under the SVJ model and for different values of β. Parameters: $r = .02, \sigma_0^2 = .05 \kappa = 2, \vartheta = .05, \rho = -.45, \zeta = .8, \lambda = 20, \mu = 0, \sigma = .03)$.

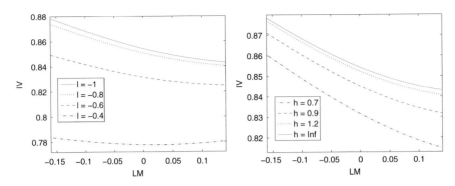

Fig. 4.17: The IVs for a 3x leveraged ETF as the floor (l, top) and cap (h, bottom) thresholds vary. Model parameters: $r = .02, \sigma_0^2 = .011$, $\kappa = 2, \vartheta = .04, \rho = -.53, \zeta = .515, \lambda = 4, \mu = 0, \sigma = .13$).

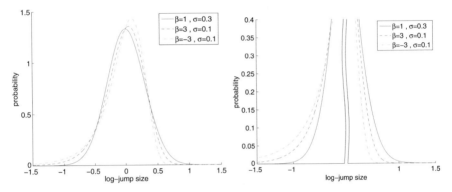

Fig. 4.18: The reference and leveraged jump distributions (top) and their tails (bottom) when Z is Gaussian, with $\mu = 0$, and for different values of β and σ.

(4.47), the left tail follows that of an exponential random variable. Therefore, for example when Z is Gaussian, the left tail of Z_L is significantly fatter, see Figure 4.18. Furthermore, even if the left tail is truncated, when $l > -1$, the likelihood of extreme negative values for Z_L can be significantly higher than that of Z. An illustrative example is displayed in Figure 4.18. On the other hand, when $\beta > 0$ and $h = \infty$, the right tail is not truncated and admits the asymptotic (4.48). Therefore, the right tail follows the same functional form as that of Z, when it is not truncated. Nevertheless, as seen in Figure 4.18, due to the form of (4.46) the likelihood of extreme positive values for Z_L may be significantly lower than that of Z.

Chapter 5
Conclusions

LETFs are unique financial products because they offer individual and institutional investors a long or short leveraged exposure with respect to a reference index or asset, without the need to borrow or rebalance dynamically by the LETF holder. Undoubtedly, these funds provide interesting trading opportunities, but they also suffer from volatility decay (see Section 2.2) and tracking errors (see Section 2.3) that may cause severe value erosion.

This book provides both empirical and theoretical studies to investigate the risk and return characteristics of LETFs and the price relationship among LETF options. Accounting for the main features of LETFs and market observations from our empirical study, we present the stochastic models for the price evolution of these funds, and propose trading strategies and risk management tools.

With the formulas provided in this book, investors can quantify the risk of any LETF and tailor the trading strategies to their specific needs. For instance, they can now identify an acceptable range of leverage ratios according to a risk measure. Once an LETF is chosen, the investor can select a trading strategy such as the stop-loss strategy in Section 3.3. Alternatively, the investor can further control the risk exposure, in terms of Delta and Vega, by constructing a portfolio of LETFs with different leverage ratios.

Our results in this book can be readily implemented to analyze various LETFs with different reference assets, but they can also serve as the building blocks for more sophisticated models and analytical tools. For instance, our explicit formulas for the admissible leverage ratio and admissible risk horizon can be useful as the benchmark for comparing similar results based on other

© The Author(s) 2016   89
T. Leung, M. Santoli, *Leveraged Exchange-Traded Funds*, SpringerBriefs
in Quantitative Finance, DOI 10.1007/978-3-319-29094-2_5

stochastic underlying price processes. Moreover, one can incorporate market frictions, such as transaction costs, into our numerical pricing procedure in Sections 4.4 and 4.7.

As the market of ETFs continues to grow in terms of market capitalization and product diversity, there are plenty of new problems for future research. In closing, let us point out a number of new directions.

On the price dynamics of LETFs, the availability of high-frequency trading data permits the analysis of the *intraday* return patterns and tracking performance of these funds compared to their reference assets. Like in our Section 2.3, one can further estimate the empirical leverage ratio of an LETF conditioned on the intraday movements of the reference asset, and understand when an LETF tends to over/under-leverage. On trading LETFs, there is a host of interesting problems. In addition to the static trading strategies studied in our Section 2.5, one can consider dynamically trading leveraged or non-leveraged ETFs. Many LETFs are referenced to an index. While the index itself is not directly tradable, there may be futures contracts written on the index. This is true for equity indexes like the S&P500, commodity indexes, and more. As such, it is also possible to replicate the LETF dynamics by trading futures (see Chapter 5 of Leung and Li (2015a)).

As for LETF options, our numerical pricing procedure can be modified to accommodate the early exercise feature that is common for American-style ETF and LETF options. While it is analytically convenient to assume continuous rebalancing, the rebalancing period in practice is often one trading day. Therefore, it is natural to examine how the rebalancing frequency effects the LETF option price, holding other features and parameters equal. For these two topics, we refer to Chapter 4 of Santoli (2015) for more details and illustrative examples. Other related issues include pricing (L)ETF options with transaction costs, dynamic discrete-time rebalancing strategies (see Avellaneda and Zhang (2010)), static hedging using vanilla options on the same reference (see Leung and Lorig (2015)), and investment with portfolio of (L)ETF options.

Leveraged exchange-traded products are also available for other reference indexes such as Nasdaq 100 and the CBOE Volatility Index (VIX), as well as other asset classes such as bonds and commodities. For many of these markets, there are vanilla and exotic derivatives, such as futures, swaps, and options, that have been liquidly traded prior to the advent of LETFs and related products. This should motivate research to investigate the connection, especially price consistency, among derivatives. To this end, it is essential to develop tractable models that capture the main characteristics of the reference underlying market as well as the ETFs and associated derivatives. Looking forward, as the market of leverage exchange-traded products becomes a sizeable connected part of the financial market, it is crucial to better understand its feedback effect and impact on systemic risk. This is important not only for individual and institutional investors, but also for regulators.

References

Ahn A, Haugh M, Jain A (2013) Consistent pricing of options on leveraged ETFs. SSRN Preprint

Avellaneda M, Lipkin M (2009) A dynamic model for hard-to-borrow stocks. Risk Mag 22(6):92–97

Avellaneda M, Zhang S (2010) Path-dependence of leveraged ETF returns. SIAM J Financ Math 1:586–603

Bates D (1996) Jumps and stochastic volatility: the exchange rate processes implicit in Deutschemark options. Rev Financ Stud 9(1):69–107

Carr P, Madan D (1999) Option pricing and the fast Fourier transform. J Comput Finance 2:61–73

Carr P, Geman H, Madan D, Yor M (2002) The fine structure of asset returns: an empirical investigation. J Bus 75(2):305–332

Cheng M, Madhavan A (2009) The dynamics of leveraged and inverse exchange-traded funds. J Invest Manag 7:43–62

Coleman T, Li Y (1994) On the convergence of reflective Newton methods for large-scale nonlinear minimization subject to bounds. Math Program 67(2):189–224

Coleman T, Li Y (1996) An interior, trust region approach for nonlinear minimization subject to bounds. SIAM J Optim 6:418–445

Cont R, Tankov P (2002) Calibration of jump-diffusion option-pricing models: a robust non-parametric approach. Working paper

Cox J, Ingersoll J, Ross S (1985) A theory of the term structure of interest rates. Econometrica 53:385–407

del Baño Rollin S, Ferreiro-Castilla A, Utzet F (2009) A new look at the Heston characteristic function. Preprint

© The Author(s) 2016
T. Leung, M. Santoli, *Leveraged Exchange-Traded Funds*, SpringerBriefs in Quantitative Finance, DOI 10.1007/978-3-319-29094-2

Deng G, Dulaney T, McCann C, Yan M (2013) Crooked volatility smiles. J Deriva Hedge Funds 19(4):278–294

Figlewski S (2010) Estimating the implied risk neutral density for the U.S. market portfolio. In: Watson M, Bollerslev T, Russell J (eds) Volatility and time series econometrics: essays in honor of Robert Engle. Oxford University Press, Oxford

Fouque J-P, Papanicolaou G, Sircar R, Sølna K (2011) Multiscale stochastic volatility for equity, interest rate, and credit derivatives. Cambridge University Press, Cambridge

Guo K, Leung T (2015) Understanding the tracking errors of commodity leveraged ETFs. In: Aid R, Ludkovski M, Sircar R (eds) Commodities, energy and environmental finance, fields institute communications. Springer, New York, pp 39–63

Hodges H (1996) Arbitrage bounds on the implied volatility strike and term structures of European-style options. J Deriv 3:23–35

Kou S (2002) A jump-diffusion model for option pricing. Manag Sci 48: 1086–1101

Lee R (2004) Option pricing by transform methods: extensions, unification, and error control. J Comput Finance 7:51–86

Leung T, Li X (2015a) Optimal mean reversion trading: mathematical analysis and practical applications. World Scientific, Singapore

Leung T, Li X (2015b) Optimal mean reversion trading with transaction costs and stop-loss exit. Int J Theor Appl Finance 18(3):1550020

Leung T, Lorig M (2015) Optimal static quadratic hedging. Working Paper

Leung T, Lorig M, Pascucci A (2014) Leveraged ETF implied volatilities from ETF dynamics. Working Paper

Leung T, Santoli M (2012) Leveraged exchange-traded funds: admissible leverage and risk horizon. J Invest Strateg 2(1):39–61

Leung T, Sircar R (2015) Implied volatility of leveraged ETF options. Appl Math Finance 22(2):162–188

Leung T, Ward B (2015) The golden target: analyzing the tracking performance of leveraged gold ETFs. Stud Econ Finance 32(3):278–297

Lord R, Fang F, Bervoets F, Oosterlee CW (2008) A fast and accurate FFT-based method for pricing early-exercise options under Lévy processes. SIAM J Sci Comput 30(4):1678–1705

Mackintosh P, Lin V (2010) Longer term plays on leveraged ETFs. Credit Suisse: Portfolio Strategy, pp 1–6

Madan D, Unal H (1998) Pricing the risks of default. Rev Deriv Res 2: 121–160

Mason C, Omprakash A, Arouna B (2010) Few strategies around leveraged ETFs. BNP Paribas Equities Derivatives Strategy, pp 1–6

Merton R (1976) Option pricing when underlying stock returns are discontinuous. J Financ Econ 3:125–144

Russell M (2009) Long-term performance and option pricing of leveraged ETFs. Senior Thesis, Princeton University

Santoli M (2015) Methods for pricing pre-earnings equity options and leveraged ETF options. PhD thesis, Columbia University

Triantafyllopoulos K, Montana G (2009) Dynamic modeling of mean-reverting spreads for statistical arbitrage. Comput Manag Sci 8:23–49

Zhang J (2010) Path dependence properties of leveraged exchange-traded funds: compounding, volatility and option pricing. PhD thesis, New York University

Index

3/2 model, 67

adjusted moneyness, 52
admissible leverage ratio, 38
admissible risk horizon, 43
authorized participants, 1

conditional value-at-risk, 41
creation, 1
cross calibration, 72

delta-neutral, 31
dual Delta matching, 77

futures portfolio, 22

gold ETFs, 22

Heston model, 65

implied dividend, 51
implied volatility, 63
intra-horizon value-at-risk, 48

leveraged benchmark, 7

moneyness scaling, 63

Put-Call Parity, 59

realized variance, 13
redemption, 1

static portfolio, 24
Stein-Stein model, 67
stochastic volatility jump-diffusion
    (SVJ), 81
stop-loss, 48
synthetic call/put, 51

take-profit level, 50
tracking performance, 22

value-at-risk, 39
volatility decay, 14

© The Author(s) 2016
T. Leung, M. Santoli, *Leveraged Exchange-Traded Funds*, SpringerBriefs
in Quantitative Finance, DOI 10.1007/978-3-319-29094-2